INORGANIC
SYNTHESES
Volume XV

Editor-in-Chief

GEORGE W. PARSHALL

Central Research Department
E. I. du Pont de Nemours & Company
Wilmington, Delaware

●●●

INORGANIC
SYNTHESES

Volume XV

McGRAW-HILL BOOK COMPANY

New York   St. Louis   San Francisco   Düsseldorf   Johannesburg
Kuala Lumpur   London   Mexico   Montreal   New Delhi
Panama   Paris   São Paulo   Singapore
Sydney   Tokyo   Toronto

To Ronald S. Nyholm
1917–1971

CONTENTS

PREFACE

Almost inevitably the content of a volume of INORGANIC SYN-THESES reflects the interests of the Editor-in-Chief to a considerable extent. Thus Volume XIII had a strong flavor of metal-metal bonded compounds and Volume XIV contains a remarkably fine collection of solid-state syntheses. The present volume is no exception. It contains syntheses of many transition-metal compounds important in homogeneous catalysis, especially complexes of olefins, hydrogen, and nitrogen, which have interested me and my colleagues. However, I hope and believe that this volume also has general interest and that the techniques described will have broad application in inorganic chemistry. One of the strengths of INORGANIC SYNTHESES is the diversity of interests among the members of the Editorial Board. Their help in assembling preparations of all sorts of inorganic compounds prevents a volume from ever becoming a collection for just the practitioners of one narrow specialty.

Perhaps the greatest strength of INORGANIC SYNTHESES is the support that it has received from inorganic chemists generally. Assembling this volume has done much to restore my faith in human nature. People have been immensely helpful in soliciting, contributing, checking, and editing syntheses. At the risk of slighting individuals by omission, I am especially indebted to Charles Van Dyke for his initiative in assembling the chapter on germanium hydride derivatives and to George Ryschkewitsch and Roland Köster for contributing the two groups of syntheses that make up most of the chapter on boron compounds. Similarly S. D. Robinson contributed and E. R. Wonchoba checked most of the chapter on triphenyl-phosphine complexes of the transition metals.

Finally it is a pleasure to thank Mrs. Kathy Jones, Mrs. Kathy Schloegel, and Mrs. Suzanne Grandel for the typing of this volume and the members of the Editorial Board for repeated editing and proofreading of it.

G. W. Parshall

NOTICE TO CONTRIBUTORS

The INORGANIC SYNTHESES series is published to provide all users of inorganic substances with detailed and foolproof procedures for the preparation of important and timely compounds. Thus the series is the concern of the entire scientific community. The Editorial Board hopes that all chemists will share in the responsibility of producing INORGANIC SYNTHESES by offering their advice and assistance in both the formulation of and the laboratory evaluation of outstanding syntheses. Help of this kind will be invaluable in achieving excellence and pertinence to current scientific interests.

There is no rigid definition of what constitutes a suitable synthesis. The major criterion by which syntheses are judged is the potential value to the scientific community. An ideal synthesis is one which presents a new or revised experimental procedure applicable to a variety of related compounds, at least one of which is critically important in current research. However, syntheses of individual compounds that are of interest or importance are also acceptable.

The Editorial Board lists the following criteria of content for submitted manuscripts. Style should conform with that of previous volumes of INORGANIC SYNTHESES. The *Introduction* should include a concise and critical summary of the available procedures for synthesis of the product in question. It should also include an estimate of the time required for the synthesis, an indication of the importance and utility of the product, and an admonition if any potential hazards are associated with the procedure. The *Procedure* should present detailed and unambiguous laboratory directions and be written so that it anticipates possible mistakes and misunderstandings on the part of the person who attempts to duplicate the procedure. Any unusual equipment or procedure should be clearly described. Line drawings should be included when they can be helpful. All safety measures should be stated clearly. Sources of unusual starting materials must be given, and, if possible, minimal standards of purity of reagents and solvents should be stated. The scale should be reasonable for normal laboratory operation, and any problems involved in scaling the procedure either up or down should be discussed. The criteria for judging the purity of the final product should be delineated clearly. The section on *Properties* should list and discuss those physical and chemical characteristics that are relevant to judging the purity of the product and to permitting its handling and use in an intelligent manner. Under *References*, all pertinent literature citations should be listed in order.

The Editorial Board determines whether submitted syntheses meet the general specifications outlined above. Every synthesis must be satisfactorily reproduced in a different laboratory from that from which it was submitted.

Each manuscript should be submitted in duplicate to the Secretary of the Editorial Board, Professor Stanley Kirschner, Department of Chemistry, Wayne State University, Detroit, Michigan, 48202, U.S.A. The manuscript should be typewritten in English. Nomenclature should be consistent and should follow the recommendations presented in "The Definitive Rules for Nomenclature of Inorganic Chemistry," *J. Am. Chem. Soc.*, **82**, 5523 (1960). Abbreviations should conform to those used in publications of the American Chemical Society, particularly *Inorganic Chemistry*.

INORGANIC
SYNTHESES
Volume XV

Chapter One

METAL COMPLEXES OF OLEFINS

Transition-metal complexes of olefins have been extensively studied as possible intermediates in homogeneous catalysis. However, in recent years, it has been recognized that olefin complexes are extremely useful intermediates for synthesis of many other transition-metal complexes. The olefin ligand is often labile and is easily replaced by a phosphine or by carbon monoxide.

The olefin complexes of iron, nickel, rhodium, and iridium described in this chapter have found broad application in the synthesis of phosphine, phosphite, and carbonyl derivatives of these metals. In Chapter Two, the synthesis of another labile olefin complex, (ethylene)bis(tricyclohexylphosphine)nickel, is described as an initial step in synthesis of a complex of dinitrogen.

The utility of the olefin complexes in synthesis has been judged sufficient to warrant their inclusion in this volume despite the hazards and difficulties involved in their preparation.

Several of the syntheses involve the reduction of a metal salt by an alkylaluminum compound. Since these reducing agents are extremely air-sensitive, even pyrophoric, extreme care is needed, and the preparation should not be undertaken without adequate facilities for handling these compounds. However, all these syntheses have been reproduced many times and give consistently good yields when conducted *exactly as described*.

1.   BIS(1,3,5,7-CYCLOOCTATETRAENE)IRON(0)

$$Fe(C_5H_7O_2)_3 + 2C_8H_8 + 3Al(C_2H_5)_3 \longrightarrow$$
$$Fe(C_8H_8)_2 + 3Al(C_2H_5)_2(C_5H_7O_2) + \tfrac{3}{2}C_2H_4 + \tfrac{3}{2}C_2H_6$$

Submitted by D. H. GERLACH* and R. A. SCHUNN†
Checked by M. A. BENNETT,‡ I. B. TOMKINS,‡ and T. W. TURNEY ‡

The synthesis of bis(1,3,5,7-cyclooctatetraene)iron(0) by reduction of anhydrous iron(III) chloride with the isopropyl Grignard reagent in the presence of cyclooctatetraene was reported recently.[1] This compound promotes oligo- and co-oligomerization of several unsaturated hydrocarbons[1] and is useful as a precursor in the preparation of a number of dihydridoiron complexes.[2] The procedure described here is simpler than the original method[1] and gives significantly higher yields.

Materials

1,3,5,7-Cyclooctatetraene§ is purified by passage through 1–2 cm. of activated alumina supported on a coarse sintered-glass filter. Anhydrous tris(2,4-pentanedionato)iron(III)¶ and triethylaluminum¶ are used without further purification.

■ **Caution.** *Triethylaluminum is a pyrophoric liquid that reacts violently with water or alcohols. It should be handled only in a rigorously oxygen- and moisture-free atmosphere, using face shield and gloves.* The solution of triethylaluminum in diethyl ether is prepared by slowly adding 50 ml. of triethylaluminum to 150 ml. of ether in a nitrogen-filled flask cooled to *ca.* 0° and is transferred to the addition funnel with a 100-ml. hypodermic syringe.

*2522 Woodridge Drive, Decatur, Ga. 30033.
†Central Research Department, E. I. du Pont de Nemours & Company, Wilmington, Del. 19898.
‡Research School of Chemistry, Australian National University, Canberra, Australia.
§Aldrich Chemical Co., 2371 North 30th St., Milwaukee, Wis. 53210.
¶Alfa Inorganics, P.O. Box 159, Beverly, Mass. 01915.

Procedure

The entire procedure is performed in an anhydrous, oxygen-free atmosphere using anhydrous, deoxygenated solvents. Standard techniques[3,4] for bench-top inert-atmosphere reactions are used in this procedure. A 1-l. three-necked flask is equipped with a magnetic stirring bar, a thermometer, a 250-ml. pressure-equalizing addition funnel with a Teflon stopcock, and a condenser topped with a nitrogen inlet connected in parallel with a mineral-oil bubbler. The flask is charged with 36 g. (0.10 mole) of tris(2,4-pentanedionato)-iron(III), 80 g. (0.77 mole) of 1,3,5,7-cyclooctatetraene, and 400 ml. of dry diethyl ether and is purged thoroughly with nitrogen. While the mixture is being cooled to − 15°*, the previously prepared solution of triethylaluminum in ether is syringed into the addition funnel. The triethylaluminum solution is added dropwise over 1.5 hours while the temperature is maintained at about − 10°. [If the temperature during the addition is allowed to rise above 0°, tris(2,4-pentanedionato)aluminum(III) forms and interferes with the purification of the product.] During the addition the red color of the mixture changes to dark brown. The mixture is stirred at − 10° for 2 hours, warmed slowly to room temperature, and gently boiled for 0.5 hour. Gas evolves during this process. The flask is cooled to − 78° for 16 hours, and the resulting black crystalline solid is collected by filtration in an inert atmosphere at low temperature.†,‡ It is washed with 100 ml. of cold (− 78°) ether and dried *in vacuo* at room temperature for several hours. The purity of the product is sufficiently high for use in the preparation of FeH_2L_4 complexes.[2] Analytically pure samples in the form of shiny black needles can be obtained by recrystallization from ether, *n*-pentane, or petroleum ether (b.p. 38–51°). The yield of the unrecrystallized product is 20 g.

---

*Dry Ice is added to a toluene bath as necessary to obtain the desired temperature. A wet ice-methanol bath may also be used but is somewhat more hazardous because of the violent reaction of triethylaluminum with water and methanol.

†Figure 1 (p.7) illustrates an apparatus which is useful for the filtration of large-scale preparations of air-sensitive compounds at temperatures down to − 78°. For smaller-scale reactions, Schlenk-tube techniques[3,4] may be used.

‡The filtrate contains highly reactive ethylaluminum compounds which react violently with water or alcohols and is most conveniently disposed of by incineration. Alternatively, the mixture may be decomposed by the careful, dropwise addition of 200 ml. of ethanol to the cooled, stirred solution (much gas is evolved), followed by the cautious addition of water.

(74%). *Anal.* Calcd. for $C_{16}H_{16}Fe$: C, 72.75; H, 6.11; Fe, 21.14. Found: C, 72.38, 72.36; H, 6.14, 6.15; Fe, 21.38, 21.33.

Properties

Bis(1,3,5,7-cyclooctatetraene)iron(0) is a black crystalline solid which is unstable toward atmospheric oxygen. It can be stored for several weeks in an inert atmosphere at room temperature without significant decomposition. Temperatures of $-40°$ or lower should be maintained if the compound is to be stored for several months.

The complex is soluble in diethyl ether and common aliphatic and aromatic solvents. It is less stable in solution and decomposes at room temperature within a few days even when a rigorously inert atmosphere is maintained. Chlorinated hydrocarbons rapidly decompose $Fe(C_8H_8)_2$, with the formation of precipitates.

The solid- and solution-state structures of $Fe(C_8H_8)_2$ have been established by x-ray crystallography[5] and infrared and proton-magnetic-resonance spectroscopy.[6] In the solid the iron atom is five-coordinate and achieves a noble-gas configuration by π bonding to six carbon atoms of one ring and four of the other.[5] In solution, $Fe(C_8H_8)_2$ is stereochemically nonrigid on the proton n.m.r. time scale. Near room temperature all ring protons are equivalent, but at low temperatures the tricoordinate and bicoordinate C_8 rings can be "frozen" out.[6]

References

1. A. Carbonaro, A. Greco, and G. Dall'Asta, *J. Organometallic Chem.*, **20**, 177 (1969).
2. D. H. Gerlach, W. G. Peet, and E. L. Muetterties, *J. Am. Chem. Soc.*, **94**, 4545 (1972).
3. D. F. Shriver, "The Manipulation of Air-Sensitive Compounds," McGraw-Hill Book Company, New York, 1969.
4. R. B. King, in "Organometallic Syntheses," J. J. Eisch and R. B. King (eds.), vol. 1, Academic Press, Inc., New York, 1965.
5. G. Allegra, A. Colombo. A. Imminzi, and I. W. Bassi, *J. Am. Chem. Soc.*, **90**, 4455 (1968).
6. A. Carbonaro, A. L. Segre, A. Greco, C. Tosi, and G. Dall'Asta, *ibid.*, **90**, 4453 (1968).

2. BIS(1,5-CYCLOOCTADIENE)NICKEL(0)

$$Ni(C_5H_7O_2)_2 + 2C_8H_{12} + 2Al(C_2H_5)_3 \longrightarrow$$

$$Ni(C_8H_{12})_2 + 2Al(C_2H_5)_2(C_5H_7O_2) + C_2H_4 + C_2H_6$$

Submitted by R. A SCHUNN*
Checked by R. BAKER,† R. J. COLE,† J. D. GILBERT,† and D. P. MADDEN†

Bis(1,5-cyclooctadiene)nickel(0) is useful for the synthesis of a variety of novel nickel complexes,[1-5] since the cyclooctadiene ligands are easily displaced. The procedure given here is based on that described by Wilke;[6] butadiene is used to prevent the formation of nickel metal.[6]

Materials

Anhydrous bis(2,4-pentanedionato)nickel(II) is obtained from $Ni(C_5H_7O_2)_2 \cdot 2H_2O$ (Alfa Inorganics) by azeotropic distillation of the water with toluene. The hot, dark-green slurry is filtered through Celite under nitrogen, and the filtrate is evaporated to dryness on a rotary evaporator. The green, solid residue is crushed in a mortar and dried at $80°/0.1 \mu m./16$ hours to give the anhydrous complex. Peroxide impurities are removed from 1,5-cyclooctadiene (Aldrich Chemicals) by filtration through grade 1 neutral alumina; the filtration is repeated until the alumina is no longer colored yellow. Further purification is not necessary, but the liquid should be used immediately. Butadiene is dried by passing it through a U trap containing Type 3A molecular sieves, condensing the gas at $-78°$, and storing it in a small stainless-steel cylinder.

■ **Caution.** *Triethylaluminum is a pyrophoric liquid that reacts violently with water or alcohols. It should be handled only in a rigorously oxygen- and moisture-free atmosphere, using face shield and gloves. The solution of triethylaluminum in toluene is most conveniently*

*Central Research Department, Experimental Station, E. I. du Pont de Nemours & Company, Wilmington, Del. 19898.
†Department of Chemistry, The University, Southampton, S09 5NH, England.

prepared in an inert-atmosphere box[7] and transferred to the addition funnel with a 100-ml. hypodermic syringe.

Procedure

The entire procedure is performed in an anhydrous, oxygen-free atmosphere using anhydrous, deoxygenated solvents. Standard techniques[7,8] for bench-top inert-atmosphere reactions are used in this procedure. A 1-l., four-necked, round-bottomed flask is equipped with a stopcock adapter having a hose end, a thermometer, and a Dry Ice condenser topped with a T tube which is connected to a mineral-oil bubbler and a source of dry nitrogen. The center neck of the flask is left open. The flask is flushed thoroughly with nitrogen and charged through the center neck with 102.8 g. (0.4 mole) of anhydrous bis(2,4-pentanedionato)nickel(II), 250 ml. of toluene, and 216 g. (2.0 moles) of 1,5-cyclooctadiene. The center neck is then equipped with a mechanical stirrer, and the mixture is stirred and cooled to $- 10°$;* the condenser is filled with a mixture of Dry Ice and acetone. Approximately 18 g. (0.33 mole) of anhydrous 1,3-butadiene is admitted to the flask through the stopcock adapter;† the gas dissolves in the cold mixture. The stopcock adapter is then replaced with a nitrogen-flushed 250-ml. pressure-equalizing dropping funnel having a Teflon stopcock, and the system is flushed with nitrogen by removing the stopper from the top of the addition funnel. A solution of 103 g. (0.9 mole) of triethylaluminum in 100 ml. of toluene is placed in the addition funnel and added dropwise to the cold, stirred mixture. The temperature is maintained at $- 10$ to $0°$; the addition requires *ca.* 45–90 minutes. During the addition, the green slurry becomes yellow-brown and a yellow crystalline solid is formed. After being stirred at $0°$ for 0.5 hour, the mixture is allowed to warm to room temperature and is stirred overnight.‡

---

*Dry Ice is added to a toluene bath as necessary to obtain the desired temperature. A wet ice-methanol bath may also be used but is somewhat more hazardous because of the violent reaction of triethylaluminum with water and methanol.

†The weight transferred to the reaction flask is determined by supporting the cylinder on a balance while conducting the gas into the reaction flask through rubber tubing attached to the hose end of the stopcock adapter.

‡Several hours of stirring are probably sufficient. Since the product is often light-sensitive, the reaction flask should be covered during this period.

The Dry Ice condenser is also allowed to warm. Gas (predominantly ethylene and ethane) is evolved.

The yellow slurry is stirred for 2–3 hours at − 15°. With rapid nitrogen flushing, the addition funnel is replaced with an adapter attached to a nitrogen line, the thermometer and mechanical stirrer are replaced with stoppers, and all joints are secured with standard-taper-joint clips (A. H. Thomas Co.). The apparatus shown in Fig. 1* is assembled and filled with nitrogen. The joints *B* and *C* are secured with standard-taper-joint clips. With nitrogen flushing through stopcock 1 and the reaction flask, the Dry Ice condenser is removed from the reaction flask and replaced by joint *A*; the joint is secured with a joint clip. Care must be taken to support the reaction flask suitably, and it must be replaced in the cold bath. The filter is im-

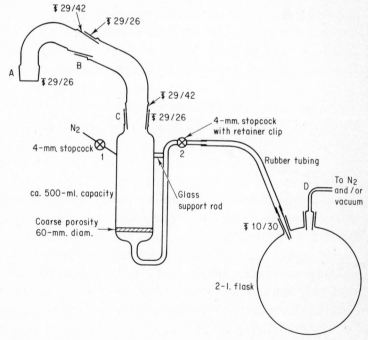

Fig. 1.   Apparatus for low-temperature filtration.

*This apparatus is useful for the filtration of large-scale preparations of air-sensitive compounds at temperatures down to − 78°. The glass adapters (joints *A*, *B*, *C*) may be replaced with a suitably adapted Flexi-joint (Cole-Parmer, Inc.). For smaller-scale reactions, Schlenk-tube techniques[7] may be used. The reaction mixture may also be filtered at room temperature, with a corresponding decrease in yield due to the increased solubility of the product.

mersed in a cold bath (− 15°), the reaction solution is transferred to the filter in portions by rotating at joint *B*, and the mixture is filtered by opening stopcock 2 and pressuring with nitrogen; the mineral-oil bubbler may be replaced by a mercury bubbler for this purpose.[7] Care must be taken to use sufficient pressure for filtering but avoid popping joints or stoppers. Alternatively, suction may be applied at *D*, but care must be taken to avoid introducing air into the system. When the entire reaction mixture has been filtered, the product is washed by adding 100 ml. of toluene to the reaction flask (via hypodermic needle through the thermometer joint), cooling it to − 15°, and pouring it onto the filter. After three washes, the filtrate is pale yellow. A final wash with 100 ml. of cold anhydrous ether facilitates drying. With nitrogen flushing rapidly through stopcock 1, the adapter at joint *C* is replaced with a stopper and stopcock 2 is closed.* A vacuum line is attacked to stopcock 2, stopcock 1 is closed, and the filter is evacuated. After 0.5 hour the filter is warmed to 25°, and after 0.5 hour the product is transferred to a vacuum-line flask and dried further (protect from light) at 25°/0.1 μm./16 hours to give 97.0 g. (89%) of $Ni(1,5-C_8H_{12})_2$; decomposes *ca.* 135–140°. *Anal.* Calcd. for $C_{16}H_{32}Ni$: C, 69.9; H, 8.8; Ni, 21.3. Found: C, 69.1; H, 8.7; Ni, 21.2.

Properties

The solid complex is decomposed after several minutes in air; solutions are more rapidly decomposed. It is moderately soluble in benzene and tetrahydrofuran, but heating these solutions above *ca.* 60° causes decomposition. It is nearly insoluble in diethyl ether and saturated hydrocarbons. The complex may be purified by extraction with benzene at 45–50°, addition of *n*-heptane to the yellow filtrate, and concentration of the mixture on a rotary evaporator.

[1]H n.m.r. (C_6D_6, ext. $(CH_3)_4Si$): δ 1.38 (s, 8, CH_2); δ 3.64 (s, bd, 4, CH).

---

*The filtrate contains highly reactive ethylaluminum compounds which react violently with water or alcohols and is most conveniently disposed of by incineration. Alternatively, the mixture may be decomposed by the careful, dropwise addition of 200 ml. of ethanol to the cooled, stirred solution (much gas is evolved), followed by the cautious addition of water.

References

1. J. Ashley-Smith, M. Green, and F. G. A. Stone, *J. Chem. Soc. (A)*, **1970**, 3161.
2. C. S. Cundy, M. Green, and F. G. A. Stone, *ibid.,* **1970**, 1647.
3. J. Browning, C. S. Cundy, M. Green, and F. G. A. Stone, *ibid.*, **1969**, 20.
4. D. H. Gerlach, A. R. Kane, G. W. Parshall, J. P. Jesson, and E. L. Muetterties, *J. Am. Chem. Soc.*, **93**, 3543 (1971).
5. U. Birkenstock, H. Bönnemann, B. Bogdanović, D. Walter, and G. Wilke, *Advan. Chem. Ser.*, **70**, 250 (1968).
6. B. Bogdanović, M. Kröner, and G. Wilke, *Annalen*, **699**, 1 (1966).
7. D. F. Shriver, "The Manipulation of Air-Sensitive Compounds," McGraw-Hill Book Company, New York, 1969.
8. R. B. King, in "Organometallic Syntheses," J. J. Eisch and R. B. King (eds.), vol. 1, Academic Press, Inc., New York, 1965.

3. ETHYLENEBIS(TRI-*o*-TOLYL PHOSPHITE)NICKEL(0) AND TRIS(TRI-*o*-TOLYL PHOSPHITE)NICKEL(0)

Submitted by W. C. SEIDEL* and L. W. GOSSER†
Checked by H. BÖNNEMANN‡ (Secs. A and B) and L. VANDE GRIEND§ and J. G. VERKADE§ (Sec. C)

Ethylenebis(tri-*o*-tolyl phosphite)nickel(0)[1,2] is a useful catalyst for many reactions of dienes.[3] It can be prepared by reduction of bis(2,4-pentanedionato)nickel(II) with triethylaluminum in the presence of excess phosphite ligand. However, much better yields are obtained by the procedure described in Sec. A, in which a stoichiometric quantity of phosphite is used in conjunction with excess ethylene.[1]

Tris(tri-*o*-tolyl phosphite)nickel(0),[4] likewise a useful catalyst,[5] is conveniently prepared by displacement of ethylene from the olefin complex by a mole of phosphite ligand as described in Sec. B. Alternatively, it can be prepared directly by reduction of hydrated nickel nitrate with sodium tetrahydroborate in acetonitrile (Sec. C). This method avoids the need to handle spontaneously flammable triethylaluminum.

*Polymer Intermediates Department, E. I. du Pont de Nemours & Company, Orange, Tex.
†Central Research Department, Experimental Station, E. I. du Pont de Nemours & Company, Wilmington, Del. 19898.
‡Max Planck Institut für Kohlenforschung, Mülheim/Ruhr, Germany.
§Department of Chemistry, Iowa State University, Ames, Iowa 50010.

A. ETHYLENEBIS(TRI-*o*-TOLYL PHOSPHITE)NICKEL(0)

$$Ni(C_5H_7O)_2 + C_2H_4 + 2P(OC_6H_4CH_3)_3 + (C_2H_5)_3Al \longrightarrow$$
$$Ni(C_2H_4)[P(OC_6H_4CH_3)_3]_2$$

Materials

This procedure can be greatly simplified if the appropriate starting materials are available. Handling of the triethylaluminum is facilitated by use of the nonpyrophoric 25 % solution in hexane available from Texas Alkyls Division of the Stauffer Chemical Co., Deer Park, Tex. 77536. (■ **Caution.** *Neat triethylaluminum is spontaneously flammable in air.*) Dehydration of the bis(2,4-pentanedionato)nickel-(II) dihydrate can be avoided by purchase of the anhydrous complex from Königswarter and Ebell, Postfach 2020, 58 Hagen/Westfalen, Germany. Tri-*o*-tolyl phosphite is available from DAP, 6140 Marienberg, Post Bensheim, Germany, or it may be prepared by the procedure of Walsh.[6]

Procedure

■ **Caution.** *The ligand, reducing agent, and product are oxygen-sensitive and must be handled under an inert atmosphere at all times.*

The synthesis is conducted in a 500-ml. two-necked flask equipped with magnetic stirrer and a gas inlet. If hydrated bis(2,4-pentane-dionato)nickel(II) (ROC/RIC, 11686 Sheldon St., Sun Valley, Calif. 91352) is to be used, 5.88 g. (20 mmoles) of the dihydrate and 125 ml. of toluene are placed in the flask, and it is fitted with a Dean-Stark trap.* The mixture is boiled under reflux under nitrogen until 0.72 ml. of water collects in the trap.

The solution of 20 mmoles of anhydrous bis(2,4-pentanedionato)-nickel in *ca.* 100 ml. of toluene obtained as above or prepared from anhydrous reagents is cooled to 0°. The Dean-Stark trap is replaced with a rubber septum, and 14.08 g. (12.4 ml., 40 mmoles) of tri-*o*-tolyl

---

*A device for separating water from an azeotropic distillate while returning toluene to the flask is available from A. H. Thomas Co., Philadelphia, Pa.

phosphite is added. The solution is bubbled with ethylene gas for 5 minutes. Then, with the ethylene stream off, 27 ml. of 25% triethyl-aluminum in hexane (0.04 mole) is added by syringe through the septum over a period of 20 minutes. The originally green solution becomes yellow. The color deepens to orange as the solution is stirred at *ca.* 0° for 3 hours.

The product is precipitated by addition of oxygen-free methanol (200 ml.) by syringe through the septum. (■ **Caution.** *The addition of methanol is accompanied by vigorous gas evolution.*) It should be added 1 ml. every 5 minutes for the first 5 ml.

The reaction mixture is transferred to an inert-atmosphere "dry-box" and cooled to − 25°. The bright yellow crystals are collected on a filter pad and recrystallized by dissolving in 35 ml. of warm (60°C.) toluene, filtering through a medium-porosity-filter frit, and precipitating by the addition of 100 ml. of cold methanol. The product is dried under high vacuum; yield 10.4 g. (66%). *Anal.* Calcd. for $C_{44}H_{46}NiO_6P_2$: C, 66.8; H, 5.8; Ni, 7.4; P, 7.8. Found: C, 66.6; H, 5.7; Ni, 7.3; P, 7.8.

Properties

The ethylene complex is a yellow, air-sensitive, crystalline solid which melts with decomposition at 118–120°.

B. TRIS(TRI-*o*-TOLYL PHOSPHITE)NICKEL(0)

$$Ni(C_2H_4)[P(OC_6H_4CH_3)_3]_2 + P(OC_6H_4CH_3)_3 \longrightarrow$$
$$Ni[P(OC_6H_4CH_3)_3]_3 + C_2H_4$$

Procedure

A solution of 10.0 g. (12.6 mmoles) of ethylenebis(tri-*o*-tolyl phosphite)nickel(0) and 4.4 g. (3.9 ml.) of tri-*o*-tolyl phosphite in 50 ml. of benzene is evaporated to dryness under vacuum. Nitrogen is admitted to the flask, and the resulting red foam is dissolved in 75 ml. of toluene. The tris(tri-*o*-tolyl phosphite)nickel is precipitated as red crystals by addition of 150 ml. of methanol and cooling; yield 12 g. (85%). The checker obtained an 88% yield of product, which gave the following analysis. *Anal.* Calcd. for $C_{63}H_{63}NiO_9P_3$: Ni, 5.3. Found: Ni, 5.3.

C. TRIS(TRI-*o*-TOLYL PHOSPHITE)NICKEL(0)

$$Ni(NO_3)_2 \cdot 6H_2O + 2NaBH_4 + 3P(OC_6H_4CH_3)_3 \longrightarrow$$
$$Ni[P(OC_6H_4CH_3)_3]_3 + 2NaNO_3 + 7H_2 + 2B(OH)_3$$

Materials

Sodium tetrahydroborate, nickel nitrate hexahydrate, and nominally anhydrous acetonitrile are commercially available. Tri-*o*-tolyl phosphite is prepared by the method of Walsh.[6]

Procedure

A solution* of 6.0 g. (20 mmoles) of nickel nitrate hexahydrate in 150 ml. of acetonitrile is prepared in a 1-l., three-necked, round-bottomed flask equipped with a nitrogen inlet, gas vent (oil bubbler), and magnetic stirrer. Tri-*o*-tolyl phosphite (30 g., 85 mmoles) is added to the stirred nickel nitrate solution. Sodium tetrahydroborate (10 g., 260 mmoles) is placed in a 125-ml. Erlenmeyer flask which is attached to one of the necks of the reaction flask with Gooch tubing as described by Fieser.[7] The flask is flushed briefly with nitrogen and surrounded with a bath of room-temperature water. The sodium tetrahydroborate is added in portions to the stirred solution. (■ **Caution.** *Vigorous gas evolution.*) The gas evolution is allowed to subside before addition of the next portion. After the addition of the sodium tetrahydroborate is complete, the pale gray-green mixture is stirred at room temperature for 2 hours. The reaction vessel is transferred to a nitrogen-filled glove box, and the reaction mixture is filtered through a medium glass frit.† The acetonitrile is

---

*Ideally the solution of nickel nitrate and tri-*o*-tolyl phosphite in acetonitrile should be prepared in a nitrogen atmosphere. In practice satisfactory results are obtained if the materials are handled in air briefly for measurement before closing and nitrogen-flushing the reaction vessel. After the addition of sodium tetrahydroborate, no oxygen can be allowed to contact the reaction mixture or the solutions containing the product. Solvents must be nitrogen-purged or degassed, with the exception of the acetonitrile for the initial reaction mixture.

†The excess sodium tetrahydroborate is collected at this point and should be handled with appropriate caution.

thoroughly evaporated from the pale yellow-green filtrate at room temperature under vacuum (oil pump), preferably in a rotary evaporator. (Complete removal of acetonitrile is necessary for satisfactory isolation of the product. The checkers found it necessary to dry the residue for 4 days.) The resulting red grease is dissolved in 150 ml. of benzene, and the solution is filtered through a medium frit. The benzene is evaporated from the filtrate under vacuum, and the red grease is triturated with 150 ml. of pentane. The pentane solution is decanted from the grease, and the grease is triturated with 150 ml. of hexane to produce red, solid tris(tri-*o*-tolyl phosphite) nickel(0), which is collected and vacuum-dried (9 g., m.p. *ca.* 125° with decomposition). The checkers obtained 9.5 g. of red-orange solid, which melted with decomposition at 140–143° and gave satisfactory elemental analyses. *Anal.* Calcd. for $C_{63}H_{63}O_9P_3Ni$: C, 67.81; H, 5.69; O, 12.90; P, 8.32; Ni, 5.26. Found: C, 67.86; H, 5.82; O, 13.21; P, 8.06; Ni, 5.02.

Properties

Tris(tri-*o*-tolyl phosphite)nickel(0) is a bright red-orange solid which decomposes above *ca.* 125–140°. The 407-nm. peak has $\varepsilon = 5.0 \times 10^3$ cm.$^{-1}$ M^{-1} and the 450-nm. shoulder has $\varepsilon = 3.7 \times 10^3$ cm.$^{-1}$ M^{-1}. The n.m.r. signal of the methyl hydrogens in free tri-*o*-tolyl phosphite is at 2.11 p.p.m. down field from internal tetramethylsilane in benzene solution, while the methyl resonance of coordinated tri-*o*-tolyl phosphite is only 1.96 p.p.m. down field.

References

1. W. C. Seidel and C. A. Tolman, *Inorg. Chem.*, **9**, 2354 (1970).
2. G. Wilke, *Angew. Chem. Int. Ed.*, **2**, 105 (1963).
3. R. G. Miller, P. A. Pinke, R. D. Stauffer, and H. J. Golden, *J. Organometallic Chem.*, **29**, C42 (1971).
4. L. W. Gosser and C. A. Tolman, *Inorg. Chem.*, **9**, 2350 (1970).
5. L. W. Gosser and G. W. Parshall, *Tetrahedron Letters*, **1971**, 2555.
6. E. N. Walsh, *J. Am. Chem. Soc.*, **81**, 3023 (1959).
7. L. F. Feiser, "Experiments in Organic Chemistry," 3d ed., p. 265, D. C. Heath and Company, Boston, 1957.

4. DI-μ-CHLOROTETRAKIS(ETHYLENE)DIRHODIUM(I), 2,4-PENTANEDIONATOBIS(ETHYLENE)RHODIUM(I), AND DI-μ-CHLOROTETRACARBONYLDIRHODIUM(I)

Submitted by RICHARD CRAMER*
Checked by J. A. MCCLEVERTY† and J. BRAY†

A variety of organic syntheses which involve reactions of olefins are catalyzed by rhodium compounds. As a consequence, considerable attention has been given to the study of the properties of olefin complexes of rhodium. The two ethylene complexes,[1,2] the preparations of which are described here, are very useful in this respect. Moreover, since ethylene is very labile and volatile, a variety of compounds (including complexes of other olefins) are easily accessible from them by nucleophilic displacement of ethylene. Displacement of ethylene by carbon monoxide is illustrated by the synthesis of di-μ-chlorotetracarbonyldirhodium(I).

No alternative preparative method for either ethylene complex has been reported. A synthesis of dichlorotetracarbonyldirhodium from "rhodium trichloride trihydrate" and carbon monoxide was described in *Inorganic Syntheses*.[3] However, it is convenient to convert $[Rh_2(C_2H_4)_4Cl_2]$ to $[Rh_2(CO)_4Cl_2]$, even though the over-all yield (60%) of the carbonyl produced in this way from "rhodium trichloride trihydrate" is less than through direct carbonylation (96%).[3]

A. DI-μ-CHLOROTETRAKIS(ETHYLENE)DIRHODIUM(I)

$$2RhCl_3 \cdot 3H_2O + 6C_2H_4 \longrightarrow$$

$$(C_2H_4)_2Rh \underset{Cl}{\overset{Cl}{\diagdown\diagup}} Rh(C_2H_4)_2 + 4HCl + 2CH_3CHO + 4H_2O$$

---

*Central Research Department, Experimental Station, E. I. du Pont de Nemours & Company, Wilmington, Del. 19898.
†Department of Chemistry, The University, Sheffield S3 7HF, England.

Procedure

"Rhodium trichloride trihydrate"* (10 g., 0.037 g. atoms of Rh) is dissolved in 15 ml. of water by warming on a steam bath. The solution is transferred to a 500-ml. round-bottomed flask containing a Teflon-covered magnetic stirring bar and 250 ml. of methanol. The flask is freed of oxygen by alternately evacuating (water pump) and repressuring with ethylene to 1 atmosphere. The methanolic solution is stirred at room temperature under ethylene at a pressure of about 1 atmosphere.† The product begins to precipitate as a finely divided solid after about an hour. It usually has the color of dichromate but occasionally is dark-rust-colored. After about 7 hours, it is collected by filtration under vacuum on a sintered-glass funnel. It is best to decant most of the liquid before transferring the solid to the filter, because the product is sometimes so finely divided that filtration is slow. One should avoid drawing air through the solid. The product is washed with about 50 ml. of methanol and dried *in vacuo* at room temperature. The yield is 4.8–5.0 g. (60–65%). *Anal.* Calcd. for C_4H_8ClRh: C, 24.70; H, 4.15; Cl, 18.23; Rh, 52.90. Found: C, 24.81; H, 4.17; Cl, 18.26; Rh, 51.23.

The product dissolves by reaction with HCl, and a second crop can be recovered by neutralizing the acid generated during synthesis. A solution of 1.5 g. of NaOH in 3 ml. of H_2O is added to the filtrate and washings from the first crop. The solution is treated with ethylene as before to recover 1.0–1.5 g. of $(C_2H_4)_2RhCl_2Rh(C_2H_4)_2$, giving a combined yield of about 6 g. or 75% of theory. Attempts to get a third crop by a second treatment with NaOH give a small amount of inferior product.

The reaction has been run successfully, using 1–30 g. of $RhCl_3 \cdot 3H_2O$ with a proportionate adjustment in the amount of solvents. Both the submitter and the checkers have found occasional samples of $RhCl_3 \cdot 3H_2O$ that gave only about half the normal yield. In these cases the yield has been improved by adding 20% aqueous NaOH to raise the initial pH of the reaction solution to 4 (pH indicator paper).

---

*Engelhard Industries, 429 Delancy St., Newark, N.J. 07105.
†Alternatively the reaction has been run by bubbling ethylene through the stirred solution at the rate of about a bubble per second.

Properties

Di-μ-chlorotetrakis(ethylene)dirhodium does not melt. The product is best characterized by elemental analysis and by its infrared spectrum (KBr wafer), which has medium to strong absorptions at 3060, 2980, 1520, 1430, 1230, 1215, 999, 952, 930, and 715 cm.$^{-1}$. It is sparingly soluble in all liquids and cannot be purified by recrystallization. The compound is relatively stable to air at room temperature, but stored samples develop the odor of acetaldehyde and darken superficially. It is preferred to store $[Rh_2Cl_2(C_2H_4)_4]$ at around $0°$.

B. 2,4-PENTANEDIONATOBIS(ETHYLENE)RHODIUM(I)

$$[Rh_2Cl_2(C_2H_4)_4] + 2CH_3COCH_2COCH_3 + 2KOH \longrightarrow$$

$$2 \quad \begin{array}{c} CH_3 \\ \diagdown \\ CH \\ \diagup \\ CH_3 \end{array} \begin{array}{c} C\!-\!O \\ \diagdown \\ \diagup \\ O\!=\!O \end{array} Rh(C_2H_4)_2 + 2KCl + 2H_2O$$

Procedure

A mixture of 5.5 g. of di-μ-chlorotetrakis(ethylene)dirhodium(I) (0.014 mole) in 50 ml. of diethyl ether and 3.0 ml. of 2,4-pentanedione (0.030 mole) is stirred under nitrogen at $-20°C$ while a solution of 10 g. of KOH in 30 ml. of water is added dropwise during 15 minutes. Stirring under N_2 is continued for 30 minutes at $-10°$. Next, 50 ml. of diethyl ether is added, and the resulting solution of product in ether is decanted from the aqueous phase through a filter. The aqueous layer is extracted five times with 25 ml. of ether. The combined extracts are filtered into the product solution, and it is cooled to $-80°$. About 4.5 g. of orange-yellow platelets crystallize (60% yield). These are separated by decanting the solvent and are dried under vacuum. The supernatant liquid is concentrated to about 250 ml. in a stream of nitrogen and is chilled to $-80°$ to get another 1.0g. of product. The combined product (5.5 g.) corresponds to a 75% yield. It melts at 144–146° (decomposes) and for many purposes does not require further purification. It can be recrystallized quickly from ether or methanol. Ethereal solutions slowly deposit a white solid if held for longer than an hour at 25°. *Anal.* Calcd. for $C_{19}H_{15}O_2$-Rh: C, 41.88; H, 5.86. Found: C, 42.08; H, 5.98.

Properties

2,4-Pentanedionatobis(ethylene)rhodium is relatively stable to air at 25° but it is best stored at 0°. It is readily soluble in a variety of organic solvents. Its infrared spectrum in a KBr wafer shows absorption at 3415 (m), 3060 (w), 2985 (w), 1575 (s), 1558 (s), 1524 (s), 1425 (m), 1372 (m), 1361 (m), 1267 (m), 1233 (m), 1221 (w), 1199 (w), 1029 (m), 1015 (w), 987 (w), 936 (m), 788 (m), and 874 (w) cm.$^{-1}$.

C. DI-μ-CHLOROTETRACARBONYLDIRHODIUM(I)

(Rhodium Carbonyl Chloride)

$$[Rh_2Cl_2(C_2H_4)_4] + 4CO \longrightarrow [Rh_2Cl_2(CO)_4] + 4C_2H_4$$

Procedure

■ **Caution.** *The reaction must be carried out in a well-ventilated hood.*

Di-μ-chlorotetrakis(ethylene)dirhodium(I) (1 g., 2.6 mmoles) is stirred at 25° with 30 ml. of diethyl ether while carbon monoxide is bubbled through the mixture at a rate of about a bubble per second. After an hour the mixture is filtered. The filtrate is concentrated to 15 ml. and chilled to 0°. Pure di-μ-chlorotetracarbonyldirhodium crystallizes as red needles (0.5 g., 50% yield); m.p. 126.5–127°. Evaporation of ether from the filtrate leaves 0.30 g. of less pure dichlorotetracarbonyldirhodium (m.p. 126–127°).

Properties[3]

Di-μ-chlorotetracarbonyldirhodium is an orange-red crystalline solid very soluble in most organic solvents (except the aliphatic hydrocarbons), producing orange solutions. The compound has a melting point of 124–125°, carbonyl stretching frequencies[4] in hexane solution at 2105 (m), 2089 (s), about 2080 (vw), 2035 (s), and 2003 (w) cm.$^{-1}$. It is quite volatile, forming a red crystalline sublimate. Although the pure compound is stable in dry air, its solutions in organic solvents decompose to insoluble brown materials when left exposed to air.

References

1. R. Cramer, *Inorg. Chem.*, **1,** 722 (1962).
2. R. Cramer, *J. Am. Chem. Soc.*, **86,** 217 (1964).
3. J. A. McCleverty and G. Wilkinson, *Inorganic Syntheses*, **8,** 211 (1966).
4. C. W. Garland and J. R. Wilt, *J. Chem. Phys.*, **36,** 1094 (1962); A. C. Yang and C. W. Garland, *J. Phys. Chem.*, **61,** 1504 (1957).

5. CYCLOOCTENE AND 1,5-CYCLOOCTADIENE COMPLEXES OF IRIDIUM(I)

Submitted by J. L. HERDE,* J. C. LAMBERT,* and C. V. SENOFF*
Checked by M. A. CUSHING†

The dimeric chloro-bridged complexes $[IrCl(1,5-C_8H_{12})]_2$ and $[IrCl(C_8H_{14})_2]_2$ have been synthesized by the reduction of an iridium(IV) salt in the presence of excess cycloolefin in aqueous ethanol and aqueous 2-propanol, respectively.[1,2] These dimeric complexes are more conveniently synthesized starting with commercially available hydrated iridium trichloride. The complexes are useful starting materials for the preparation of the complexes $IrCl[P(C_6H_5)_3]_3$ and $IrCl[As(C_6H_5)_3]_3$,[3] as well as a series of four- and five-coordinate iridium(I) cations containing phosphite, phosphine, and arsine ligands.[4]

A. DI-μ-CHLORO-BIS(1,5-CYCLOOCTADIENE)DIIRIDIUM(I)

$$2IrCl_3 + 2C_8H_{12} + 2CH_3CH_2OH \longrightarrow$$

$$Ir_2Cl_2(C_8H_{12})_2 + 4HCl + 2CH_3CHO$$

*Department of Chemistry, University of Guelph, Guelph, Ontario, Canada.
†Central Research Department, E. I. du Pont de Nemours & Company, Wilmington, Del. 19898.

Procedure

A 150-ml., three-necked, round-bottomed flask containing a magnetic stirring bar is charged with 2.0 g. of iridium trichloride hydrate,* 34 ml. of 95% ethanol, 17 ml. of water, and 6.0 ml. of 1,5-cyclooctadiene.† One neck of the flask is equipped with an inlet for nitrogen, another neck is equipped with a water-jacketed condenser, and the third neck is equipped with a nitrogen-bubbling tube. A slow stream of nitrogen is passed through the system, and the solution is stirred and refluxed for 24 hours, during which time the brick-red product precipitates from the solution. The mixture is allowed to cool to room temperature and di-μ-chloro-bis(1,5-cyclooctadiene)diiridium(I) is collected by filtration, washed with ice-cold methanol to remove the last traces of unreacted 1,5-cyclooctadiene, and then dried *in vacuo* at 25° for 8 hours. Yield is 1.5 g. (72%), m.p. > 200° (decomposes). *Anal.* Calcd. for $C_{16}H_{24}Cl_2Ir_2$: C, 28.61; H, 3.60; Cl, 10.56; Ir, 57.23. Found: C, 28.68; H, 3.64; Cl, 10.51; Ir, 57.17.

Properties

Di-μ-chloro-bis(1,5-cyclooctadiene)diiridium(I) is an orange-red, air-stable solid. It is soluble in chloroform and benzene, somewhat less soluble in acetone, and insoluble in ether. Other physical properties of the complex are given in the literature.[1]

B. DI-μ-CHLOROTETRAKIS(CYCLOOCTENE)DIIRIDIUM(I)

$$2IrCl_3 + 4C_8H_{14} + 2(CH_3)_2CH(OH) \longrightarrow$$
$$Ir_2Cl_2(C_8H_{14})_4 + 4HCl + 2(CH_3)_2CO$$

Procedure

A 150-ml., three-necked, round-bottomed flask is charged with 2.0 g. of hydrated iridium trichloride, 22 ml. of 2-propanol (isopropyl

---

*A suitable starting material with the approximate composition $IrCl_3 \cdot 3H_2O$ is available from Engelhard Industries, 429 Delancy St., Newark, N.J. 07105.

†1,5-Cyclooctadiene obtained from the Columbian Carbon Co., Inc., Princeton, N.J., may be used without further purification.

alcohol), 8 ml. of water, and 4.0 ml. of cyclooctene.* One neck of the flask is equipped with an inlet for nitrogen, another neck is equipped with a water-jacketed condenser, and the third neck is equipped with a nitrogen-bubbling tube. A slow stream of nitrogen is passed through the system, and the solution is stirred well at room temperature for a few minutes by means of a magnetic stirring bar. The solution is refluxed at 78° for 3 hours, during which time it changes from dark red to orange-yellow and di-μ-chlorotetrakis-(cyclooctene)diiridium(I) precipitates from the solution. The mixture is allowed to cool to room temperature, and di-μ-chlorotetrakis-(cyclooctene)diiridium(I) is collected *quickly* by filtration and washed *quickly* with ice-cold methanol to remove the last traces of unreacted cyclooctene. (The filtration and washing can be done better in an inert-atmosphere box if one is available.) After drying *in vacuo* at 25° for 4 hours, the yield is 1.5 g. (59%). M.p. 150° (decomposes). *Anal.* Calcd. for $C_{32}H_{56}Cl_2Ir_2$: C, 42.92; H, 6.25; Cl, 7.85; Ir, 42.98. Found: C, 42.80; H, 6.50; Cl, 7.88; Ir, 42.86.

Properties

Di-μ-chlorotetrakis(cyclooctene)diiridium(I) is an air-sensitive yellow solid. Upon exposure to air, it readily decomposes to a dark-green solid which eventually blackens. It is soluble in benzene, chloroform, and acetone; solutions in these solvents are extremely sensitive to oxygen. The complex may be stored for a prolonged period of time *in vacuo* in a vacuum desiccator. Other physical properties of the complex are given in the literature.[5]

References

1. G. Winkhaus and H. Singer, *Chem. Ber.*, **99**, 3610 (1966).
2. A. van der Ent and A. L. Onderdelinden, *Inorganic Syntheses*, **14**, 92 (1973).
3. M. A. Bennett and D. L. Milner, *J. Am. Chem. Soc.*, **91**, 6983 (1969).
4. L. M. Haines and E. Singleton, *J. Chem. Soc., Dalton Trans.*, **1972**, 1891.
5. J. L. Herde and C. V. Senoff, *Inorg. Nucl. Chem. Letters*, **7**, 1029 (1971).

*Cyclooctene obtained from the Columbian Carbon Co., Inc., Princeton, N.J., may be used without further purification.

Chapter Two

METAL COMPLEXES OF DINITROGEN AND OF HYDROGEN

6. trans-(DINITROGEN)BIS[ETHYLENEBIS-(DIETHYLPHOSPHINE)]HYDRIDOIRON(II) TETRAPHENYLBORATE

Submitted by M. J. MAYS* and B. E. PRATER*
Checked by E. R. WONCHOBA† and G. W. PARSHALL†

Although dinitrogen complexes of ruthenium and osmium have been prepared in considerable numbers, few well-characterized iron compounds containing dinitrogen are known.[1-3] Herein is described the synthesis of trans-[FeH(N$_2$){(C$_2$H$_5$)$_2$ PCH$_2$CH$_2$P(C$_2$H$_5$)$_2$}$_2$]-[B(C$_6$H$_5$)$_4$] from trans-[FeHCl{(C$_2$H$_5$)$_2$ PCH$_2$CH$_2$P(C$_2$H$_5$)$_2$}$_2$], the reaction involving displacement of chloride ion by dinitrogen. The preparations of the chlorohydride and of its precursor trans-[FeCl$_2${(C$_2$H$_5$)$_2$ PCH$_2$CH$_2$P(C$_2$H$_5$)$_2$}$_2$] are modifications of those originally developed by Chatt and Hayter,[4] who first prepared these compounds. The general synthetic method is also applicable to preparation of analogous dinitrogen complexes of ruthenium and osmium.[3]

*University Chemical Laboratory, Lensfield Road, Cambridge, CB2 IEW, England.
†Central Research Department, E. I. du Pont de Nemours & Company, Wilmington, Del. 19898.

A. *trans*-DICHLOROBIS-
[ETHYLENEBIS(DIETHYLPHOSPHINE)]IRON(II)

$$FeCl_2 + 2(C_2H_5)_2PCH_2CH_2P(C_2H_5)_2 \longrightarrow$$

$$[FeCl_2\{(C_2H_5)_2PCH_2CH_2P(C_2H_5)_2\}_2]$$

Procedure

■ **Caution.** *Ethylenebis(diethylphosphine) is toxic and spontaneously flammable.* It must be handled in a well-ventilated area with complete exclusion of oxygen. Iron(II) chloride and the product are moisture-sensitive, decomposing rapidly in water and slowly in moist air. All solvents should be dried, e.g., over molecular sieves, before use.

A 500-ml. two-necked flask, fitted with a nitrogen inlet and outlet, and a reflux condenser, is flushed with nitrogen and then charged with anhydrous iron(II) chloride* (1.25 g., 9.8 mmoles) and 60 ml. of benzene. A solution of ethylenebis(diethylphosphine)† (4.05 g., 19.7 mmoles) dissolved in oxygen-free benzene (40 ml.) is added, producing an immediate bright green color. The mixture is boiled under reflux for 75 minutes, and the benzene is removed *in vacuo*. The flask is warmed if necessary to prevent solidification of the benzene. The bright green solid which remains is recrystallized by dissolving in boiling hexane or petroleum ether (b.p. 60–80°, *ca*. 600 ml.) under nitrogen, decanting the solution from any residue, and allowing it to cool. The product is obtained as green needles, which are filtered, washed with cold petroleum ether, and air-dried. The yield is 3.2–3.8 g. (61–72%).

Properties[4]

trans-Dichlorobis[ethylenebis(diethylphosphine)]iron(II) crystallizes from petroleum ether as light-green needles, m.p. 168–171.5° (decomposes). It is soluble in nonpolar solvents and is very moisture-sensitive in solution.

---

*Convenient preparations are described on p. 39 and in *Inorganic Syntheses*, **6**, 172 (1960).
†Orgmet, 300 Neck Road, Haverhill, Mass. 01830.

B. *trans*-CHLOROBIS-
[ETHYLENEBIS(DIETHYLPHOSPHINE)]HYDRIDOIRON(II)

$$4[FeCl_2\{(C_2H_5)_2PCH_2CH_2P(C_2H_5)_2\}_2] + LiAlH_4 \longrightarrow$$
$$4[FeClH\{(C_2H_5)_2PCH_2CH_2P(C_2H_5)_2\}_2] + LiAlCl_4$$

Procedure

■ **Caution.** *Lithium tetrahydroaluminate and the product are very air-sensitive and must be handled in an inert atmosphere. Solvents should be rigorously anhydrous.*

A 500-ml. three-necked flask containing a magnetic stirring bar is fitted with a nitrogen inlet and outlet and a pressure-equalized dropping funnel. After flushing with nitrogen, the flask is charged with *trans*-dichlorobis[ethylenebis(diethylphosphine)]iron(II), (2.00 g., 3.7 mmoles) and 60 ml. of anhydrous tetrahydrofuran. The complex dissolves on stirring to give a green solution, which is treated dropwise with a suspension of lithium tetrahydroaluminate in dry tetrahydrofuran. As the reaction proceeds, the mixture becomes intense red-brown and then pale yellow, at which point the addition is stopped.* The LiAlH$_4$ suspension in the dropping funnel is replaced by absolute ethanol (20 ml.), which is added dropwise to the mixture to destroy any excess LiAlH$_4$ present. (■ **Caution.** *This addition causes vigorous effervescence and the evolution of a large volume of hydrogen.*) The red-brown color is restored in this process, but the color restoration may take up to one hour. The solvent is removed in a fast stream of nitrogen, and the orange-red residue is extracted with 50-ml. portions of oxygen-free hexane or petroleum ether (b.p. 60–80°) until the extracts are almost colorless. The orange-red solution is centrifuged to remove suspended lithium and aluminum salts and is evaporated to dryness *in vacuo* to give orange-red crystals of product. Yield is *ca.* 1.55 g. (83 %).

The chlorohydride so obtained is sufficiently pure for the preparation of the dinitrogen complex, but it can be further purified by recrystallization from petroleum ether (b.p. 60–80°).[4]

---

*An excess gives a green color but, in the checker's experience, does not seriously reduce the yield.

Properties[4]

trans-Chlorobis[ethylenebis(diethylphosphine)]hydridoiron(II) crystallizes from petroleum ether as red needles, m.p. (*in vacuo*) 154.5–155.5° (decomposes). The infrared spectrum (Nujol mull) shows v_{Fe-H} at 1849 cm.$^{-1}$. The Fe—H absorption in the ^{1}H n.m.r. spectrum appears at 39.1 p.p.m. high field from H_2O (external standard). The complex is soluble in nonpolar organic solvents and is very oxygen-sensitive.

C. *trans*-(DINITROGEN)BIS [ETHYLENEBIS-(DIETHYLPHOSPHINE)]HYDRIDOIRON(II) TETRAPHENYLBORATE

$$[FeHCl\{(C_2H_5)_2PCH_2CH_2P(C_2H_5)_2\}_2] + Na[B(C_6H_5)_4] \xrightarrow{N_2}$$

$$[FeH(N_2)\{(C_2H_5)_2PCH_2CH_2P(C_2H_5)_2\}_2][B(C_6H_5)_4] + NaCl$$

Procedure

■ **Caution.** *Both the starting iron complex and the product are very oxygen-sensitive in solution. Solutions should be handled in an inert atmosphere and should be prepared from nitrogen-flushed solvents.*

A solution of *trans*-chlorobis[ethylenebis(diethylphosphine)]-hydridoiron(II) (1.55 g., 3.03 mmoles) in 120 ml. of acetone is stirred magnetically under nitrogen in a 500-ml. flask. Sodium tetraphenyl-borate (1.04 g., 3.03 mmoles) dissolved in 50 ml. of oxygen-free acetone is added, and the mixture is stirred for 15 minutes. During this time it becomes pale and a fine precipitate of sodium chloride is produced. On cooling, orange crystals of product separate slowly* and are filtered, washed with petroleum ether, and air-dried for a short time. Yield is 1.40–1.55 g. (57–63%). The compound can be purified by dissolving in acetone, centrifuging, and adding petroleum ether to the filtrate to give orange needles. *Anal.* Calcd. for $C_{44}H_{69}BFeN_2P_4$: C, 64.7; H, 8.4; N, 3.4. Found: C, 64.7; H, 8.3; N, 3.2.

---

*These may be preceded by a small amount of an amorphous, dark material which should be removed by centrifugation or filtration.

Properties

The crystalline complex is moderately stable in air for short periods (*ca.* 2 hours) and is apparently unchanged after 2 years when stored under nitrogen in a tightly stoppered container. In solution it is considerably more air-sensitive.

Its infrared spectrum (Nujol mull) shows a strong, very sharp absorption at 2090 cm.$^{-1}$, which is assigned to $v(N\equiv N)$. A high-field quintet in its ^1H n.m.r. spectrum at τ 28.2 [in acetone-d$_6$ solution, $(CH_3)_4Si$ internal standard], with a coupling constant $J(PH) = 49$ Hz., confirms the trans configuration of the complex.[2,3]

References

1. A. Sacco and M. Aresta, *Chem. Commun.* **1968,** 1223.
2. G. M. Bancroft, M. J. Mays, and B. E. Prater, *ibid.*, **1969,** 585.
3. G. M. Bancroft, M. J. Mays, E. Prater, and F. P. Stefanini, *J. Chem. Soc. (A)*, **1970,** 2146.
4. J. Chatt and R. G. Hayter, *J. Chem. Soc.*, **1961,** 5507.

7. *trans*-BIS(DINITROGEN)BIS[ETHYLENEBIS- (DIPHENYLPHOSPHINE)]MOLYBDENUM(0)

$$Mo(acac)_3{}^* + 2(C_6H_5)_2PCH_2CH_2P(C_6H_5)_2 + Al(C_2H_5)_3 + 2N_2 \longrightarrow$$

$$[Mo(N_2)_2\{(C_6H_5)_2PCH_2CH_2P(C_6H_5)_2\}_2] + Al(acac)_3 + C_2H_4 + C_2H_6$$

Submitted by M. HIDAI,† K. TOMINARI,† Y. UCHIDA,† and A. MISONO†
Checked by L. K. ATKINSON,‡ A. H. MAWBY,‡ and D. C. SMITH‡

After Allen's discovery of ruthenium nitrogen complexes,[1] many nitrogen complexes of the transition metals have been prepared by various methods. Molybdenum[2] and iron[3] nitrogen complexes prepared recently are exceptionally interesting because of the key

*acac = acetylacetonato (2, 4-pentanedionato).

†Department of Industrial Chemistry, The University of Tokyo, Hongo, Tokyo, Japan.
‡Petrochemical and Polymer Laboratory, Imperial Chemical Industries, Ltd., The Heath, Runcorn, Cheshire, England.

role of these metals in both chemical and biological nitrogen fixation. *trans*-Bis(dinitrogen)bis[ethylenebis(diphenylphosphine)] molybdenum(0) is obtained by the reaction between tris(2,4-pentanedionato)-molybdenum(III), an organoaluminum compound, ethylenebis-(diphenylphosphine), and nitrogen gas as described here. The bis-(dinitrogen) complex is also accessible by zinc reduction of $MoOCl_2$-$[(C_6H_5)_2PC_2H_4P(C_6H_5)_2](C_4H_8O)$ and by sodium amalgam reduction of $MoCl_2[(C_6H_5)_2PC_2H_4P(C_6H_5)_2]_2.$[4]

Procedure

■ **Caution.** *Triethylaluminum is spontaneously flammable and must be handled in an inert atmosphere.**

All operations are carried out under a nitrogen atmosphere. Before the reaction is begun, a 200-ml. four-necked flask *A* (Fig. 2) containing a magnetic stirring bar is evacuated via a vacuum pump

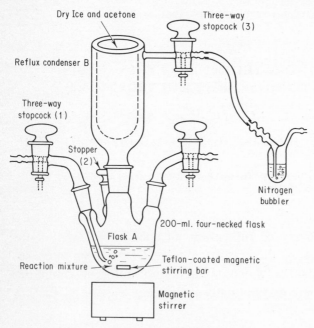

Fig. 2. Inert-atmosphere reaction apparatus.

*A nonpyrophoric solution of 25% triethylaluminum in toluene convenient for use in this synthesis is available from Texas Alkyls, Inc., P.O. Box 600, Deer Park, Tex. 77536.

and refilled with dry prepurified nitrogen.* This process is repeated at least three times. Flask *A* is charged with 4.0 g. (10.2 mmoles) of tris(2,4-pentanedionato)molybdenum(III),[5] 8.1 g. (20.4 mmoles) of ethylenebis(diphenylphosphine) [1,2-bis(diphenylphosphino)ethane]†,[6] and 60 ml. of toluene* under a nitrogen atmosphere. The flask is fitted with a reflux condenser *B* cooled with Dry Ice and acetone in a countercurrent of nitrogen. The mixture in the flask is cooled to a temperature of *ca.* − 20°, and 8.3 ml. (60.4 mmoles) of triethylaluminum is injected by means of a hypodermic syringe through stopper opening 2. The mixture is stirred with a magnetic stirrer while nitrogen gas is bubbled through the mixture from stopcock 1, escaping through stopcock 3. The temperature is held at *ca.* − 20° for the first half an hour, then gradually raised to room temperature and stirred for 4 or 5 days. The nitrogen flow is continued through this period. The mixture becomes reddish-brown, and some of the orange nitrogen complex precipitates. The reflux condenser *B* is now removed from flask *A*. When 25 ml. of hexane is added to the solution, an additional quantity of the orange complex is deposited. The solution is filtered under a nitrogen atmosphere by the filtration apparatus described in Fig. 3. Receiver flask *C*, previously filled with nitrogen gas, is connected to flask *A* through a glass-tube bridge fitted with a medium-porosity glass filter. The

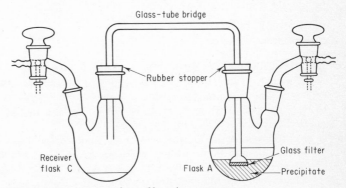

Fig. 3. Inert-atmosphere filtration apparatus.

---

*Toluene and hexane are purified by distillation from sodium wire in an atmosphere of nitrogen. The nitrogen gas is purified by passing it through an activated-copper column at a temperature of 170°.

†Available from Alfa Products, P.O. Box 159, Beverly, Mass. 01915.

nitrogen pressure in flask *A* is held a little higher than 1 atmosphere, while that in flask *C* is kept at 1 atmosphere. The difference of pressure between flask *A* and flask *C* allows the solution to be filtered. After the filtration is finished, the precipitate is washed three times with hexane and dried *in vacuo*. A 1.3-g. yield (13 %) of the orange complex is obtained.* The crude complex can be recrystallized under nitrogen in a Schlenk apparatus,[7] using the following procedure. Pure dry toluene is added from a syringe, inserted through a stopcock, until all the precipitate *just dissolves*. If hexane is added to this solution through the stopcock, well-shaped orange crystals will be obtained. *Anal.* Calcd. for $C_{52}H_{48}N_4P_4Mo$: C, 65.83; H, 5.10; N, 5.91. Found: C, 65.35; H, 5.38; N, 6.31.

Properties

The complex is an orange crystalline solid and fairly stable in air. It decomposes above *ca.* 150° with evolution of the theoretical amount of nitrogen gas. It is soluble in toluene and tetrahydrofuran, but insoluble in petroleum ether. A very strong band at 1970 cm.$^{-1}$ in the infrared spectrum as well as a very weak band at 2020 cm.$^{-1}$ is assignable to the $N{\equiv}N$ stretching mode of the *trans*-coordinated nitrogen molecules. The nitrogen is irreversibly displaced by hydrogen and by carbon monoxide.

References

1. A. D. Allen and C. V. Senoff, *Chem. Commun.*, **1965**, 621.
2. M. Hidai, K. Tominari, Y. Uchida, and A. Misono, *ibid.*, **1969**, 814, 1392; *J. Am. Chem. Soc.*, **94**, 110 (1972).
3. C. H. Campbell, A. R. Dias, M. L. H. Green, T. Saito, and M. G. Swanwyck, *J. Organometallic Chem.*, **14**, 349 (1968).
4. L. K. Atkinson, A. H. Mawby, and D. C. Smith, *Chem. Commun.*, **1971**, 157.
5. M. L. Larson and F. W. Moore, *Inorganic Syntheses* **8**, 153 (1966).
6. W. Hewertson and H. R. Watson, *J. Chem. Soc.*, **1962**, 1490.
7. D. F. Shriver, "The Manipulation of Air-sensitive Compounds," McGraw-Hill Book Company, New York, 1969.

---

*The checkers obtained a yield of 11.2 % (average of six experiments). Their best yields were obtained by precipitating the product from toluene with diethyl ether in a simple Schlenk apparatus. The complex precipitated in this way gave satisfactory analyses but showed an additional $N{\equiv}N$ stretching absorption at 1985 cm.$^{-1}$, which is believed to arise from crystal-lattice effects.[4]

8. (μ-DINITROGEN-*N,N'*)BIS-
[BIS(TRICYCLOHEXYLPHOSPHINE)NICKEL(0)]

Submitted by P. W. JOLLY* and K. JONAS*
Checked by W. H. KNOTH†

(μ-Dinitrogen-*N*, *N'*)bis[bis(tricyclohexylphosphine)nickel] has been prepared by two routes: (1) treatment of bis(2,4-pentanedionato)-nickel with trimethylaluminum in the presence of tricyclohexyl-phosphine and nitrogen[1,2] and (2) substitution of the ethylene molecule in (ethylene)bis(tricyclohexylphosphine)nickel by tricyclo-hexylphosphine followed by reaction with nitrogen.[2] The latter method is described here because it avoids the use of the pyrophoric trimethylaluminum. The results are very sensitive to the experimental conditions, and the procedure described below must be followed exactly.

A. (ETHYLENE)BIS(TRICYCLOHEXYLPHOSPHINE)NICKEL[3]

$$Ni(C_5H_7O_2)_2 + 2P(C_6H_{11})_3 + C_2H_4 + (C_2H_5)_2AlOC_2H_5 \longrightarrow$$
$$Ni(C_2H_4)[P(C_6H_{11})_3]_2 + Al(OC_2H_5)(C_5H_7O_2)_2 + C_2H_6 + C_2H_4$$

Procedure

■ **Caution.** *Alkylaluminum compounds, tricyclohexylphosphine, and the products are extremely air-sensitive and must be handled in a rigorously oxygen-free atmosphere.*

Tricyclohexylphosphine (9.7 g., 35 mmoles) and 4.45 g. (17.3 mmoles) of anhydrous bis(2, 4-pentanedionato)nickel(II)[4] are added to a 250-ml. three-necked flask fitted with a gas inlet tube, 100-ml. dropping funnel (with pressure-equalizing side arm), and a ground-glass joint with top (for evacuating purposes). The apparatus is evacuated and is filled with argon. Sufficient anhydrous benzene (*ca.* 15 ml.) to dissolve the reactants is added, followed by 70 ml. of anhydrous ether. The solution is stirred magnetically while a slow stream of ethylene is bubbled through the solution. Ethoxydiethyl-

*Max Planck Institut für Kohlenforschung, Mülheim/Ruhr, Germany.
†Central Research Department, E. I. du Pont de Nemours & Company, Wilmington, Del. 19898.

aluminum* (6.5 ml.) in 30 ml. of ether is added from a dropping funnel over a period of 30 minutes. The color of the reaction mixture changes from green to yellow-brown. The mixture is stirred overnight, and the precipitated bright yellow (ethylene)bis(tricyclohexylphosphine)nickel(0) complex is collected on a filter pad under argon, washed with ether (2 × 25 ml.), and dried under high vacuum. The compound as prepared is analytically pure (yield 7.1 g., 63%).

B. (μ-DINITROGEN-N, N')BIS-[BIS(TRICYCLOHEXYLPHOSPHINE)NICKEL(0)]

$$Ni(C_2H_4)[P(C_6H_{11})_3]_2 + P(C_6H_{11})_3 \longrightarrow Ni[P(C_6H_{11})_3]_3 + C_2H_4$$

$$2Ni[P(C_6H_{11})_3]_3 + N_2 \longrightarrow [(C_6H_{11})_3P]_2Ni\text{-}N_2\text{-}Ni[P(C_6H_{11})_3]_2$$

Procedure

■ **Caution.** *Tricyclohexylphosphine and all the nickel compounds used in this preparation are extremely air-sensitive and must be handled with rigorous exclusion of oxygen.* A solution of 20.4 g. (31.5 mmoles) of (ethylene)bis(tricyclohexylphosphine)nickel and 8.8 g. (31.5 mmoles) of tricyclohexylphosphine is prepared in 1.5 l. of anhydrous, oxygen-free toluene. The reaction mixture is heated at 100° for 3 hours while argon is bubbled through the mixture by means of a glass sintered disk. The solution is evaporated to less than 100 ml. volume. Oxygen-free nitrogen (obtainable by bubbling through ethoxydiethylaluminum) is passed through the solution for one hour to give a dark precipitate of 6.9 g. (34%) of the desired dinitrogen complex. The product may be recrystallized from toluene under nitrogen. *Anal.* Calcd. for $C_{72}H_{132}N_2Ni_2P_4$: Ni, 9.3%; N_2, 22.4 ml./mmole. Found: Ni, 9.4%; N_2, 22.9 ml./mmole.†

Properties

The presence of a linear Ni—N—N—Ni system in this complex has been confirmed by an x-ray structural determination.[2] A cryo-

---

*The checker found that triethylaluminum worked better. Both alkylaluminum compounds are available from Texas Alkyls, Inc., P.O. Box 600, Deer Park, Tex. 77536.

†Nitrogen measured by displacement with 1,5,9-cyclododecatriene in benzene; gas identified by mass spectrometry as pure N_2, volume corrected to S.T.P.

scopic molecular-weight determination in benzene shows that the (dinitrogen)bis[nickel] complex is dissociated in solution. This phenomenon also accounts for the observation of a band at 2028 cm.$^{-1}$ (toluene) in the infrared spectrum, which is attributable to the $N = N$ stretching frequency of (μ-dinitrogen-N,N')bis(tricyclohexylphosphine)nickel.

The nitrogen molecule is readily displaced from solutions of the complex by argon to give the co-ordinatively unsaturated bis-(tricyclohexylphosphine)nickel complex, $Ni[P(C_6H_{11})_3]_2$ in solution. This complex readily undergoes a variety of co-ordinative-addition and co-ordinative-oxidation reactions.[2,5]

References

1. P. W. Jolly and K. Jonas, *Angew. Chem.*, **80,** 705 (1968).
2. P. W. Jolly, K. Jonas, C. Kruger, and Y.-H. Tsay, *J. Organometallic Chem.*, **33,** 109 (1971).
3. G. Herrmann, dissertation, Technische Hochschule, Aachen, 1963.
4. R. A. Schunn, *Inorganic Syntheses*, **15,** 5 (1974).
5. K. Jonas and G. Wilke, *Angew. Chem.*, **81,** 534 (1969).

9. DINITROGEN AND HYDROGEN COMPLEXES OF RUTHENIUM

$$RuHCl[(C_6H_5)_3P]_3 + N_2 \xrightarrow{Et_3Al} RuH_2(N_2)[(C_6H_5)_3P]_3$$

$$RuHCl[(C_6H_5)_3P]_3 + H_2 \xrightarrow{Et_3Al} RuH_4[(C_6H_5)_3P]_3$$

Submitted by W. H. KNOTH*
Checked by E. HOEL† and M. F. HAWTHORNE†

The original reports related to $RuH_2(N_2)[(C_6H_5)_3P]_3$ concerned spectroscopic studies on nitrogen-saturated solutions of $RuH_2[(C_6H_5)_3P]_4$.[1] It was evident that a nitrogen complex had formed, and it was speculated that this arose by the combination of molecular

*Central Research Department, Experimental Station, E. I. du Pont de Nemours & Company, Wilmington, Del. 19898.
†Department of Chemistry, University of California, Los Angeles, Calif. 90024.

nitrogen with a species formed by dissociation of triphenylphosphine from $RuH_2[(C_6H_5)_3P]_4$. Attempts to isolate the nitrogen complex from these solutions were unsuccessful because of interference from the dissociated triphenylphosphine. A successful synthesis[2] of $RuH_2(N_2)[(C_6H_5)_3P]_3$ was achieved using $RuHCl[(C_6H_5)_3P]_3$ as a starting material; this route avoids the problems caused by excessive triphenylphosphine and is the route described here.

Procedure

A. DINITROGENDIHYDRIDOTRIS(TRIPHENYLPHOSPHINE)-RUTHENIUM(II)

■ **Caution.** *Triethylaluminum ignites spontaneously in air*.

A three-necked 500-ml. flask is thoroughly dried, flushed with nitrogen, and fitted with a magnetic stirrer, a gas inlet tube, a serum stopper, and a condenser cooled with a Dry Ice–acetone mixture. Ether (300 ml.) and chlorohydridotris(triphenylphosphine)ruthenium (II)(toluene solvate,[3] 9.8 g., 9.7 mmoles) are placed in the flask. The gas inlet tube is adjusted so that it ends just above the surface of the stirred reaction mixture, and a moderate stream of nitrogen is passed through the tube. Triethylaluminum (8 ml.) is added in one portion by injection through the serum stopper, and the mixture is stirred for 4 hours. It may be necessary to add more ether if excessive evaporation occurs.

A nitrogen atmosphere is maintained throughout the following steps either by operation in a dry-box or by Schlenk-tube techniques.[4] The reaction mixture is filtered. The slightly off-white solid thus obtained is washed with ether and dried in vacuum at ambient temperature to obtain 7.6 g. (85%) of $RuH_2(N_2)[(C_6H_5)_3P]_3$. This crude product may be recrystallized by dissolution in benzene (100 ml.) with stirring for at least 30 minutes at ambient temperature, followed by filtration and the addition of hexane (400 ml.) to the filtrate. Crystalline $RuH_2(N_2)[(C_6H_5)_3P]_3$ (5.8 g., 76% recovery) separates slowly (16 hours) as a slightly deeper colored solid than the unrecrystallized product. *Anal.* Calcd. for $C_{54}H_{47}N_2P_3Ru$: C, 70.7; H, 5.2; N, 3.1; P, 10.1; Ru, 10.9. Found: C, 70.9; H, 5.3; N, 3.2; P, 10.2; Ru, 10.2.

B. TETRAHYDRIDOTRIS(TRIPHENYLPHOSPHINE)-
   RUTHENIUM(IV)

■ **Caution.** *Triethylaluminum ignites spontaneously in air.*

A three-necked 500-ml. flask is thoroughly dried, flushed with hydrogen, and fitted with a magnetic stirrer, a gas inlet tube, a serum stopper, and a condenser cooled with a solid carbon dioxide–acetone mixture. Ether (300 ml.) and $RuHCl[(C_6H_5)_3P]_3 \cdot C_6H_5\text{-}CH_3{}^3$ (6.8 g., 6.7 mmoles) are placed in the flask. The gas inlet tube is adjusted so it ends just above the surface of the stirred reaction mixture, and a moderate stream of hydrogen is passed through the tube. Triethylaluminum (6 ml.) is added by injection through the serum stopper, and the mixture is stirred for 3 hours. The mixture is filtered in an argon atmosphere to obtain 4–5 g. (67–84%) of $RuH_4[(C_6H_5)_3P]_3$. The crude product may be recrystallized as follows: 3 g. is placed in toluene (75 ml.) through which hydrogen is bubbled continuously. The mixture is heated slowly to 65° to obtain a clear dark-red solution which is allowed to cool while the hydrogen stream is maintained. Crystalline $RuH_4[(C_6H_5)_3P]_3$ (1.4 g., 47% recovery) separates as a light-tan solid. *Anal.* Calcd. for $C_{54}H_{49}P_3Ru$: C, 72.8; H, 5.5. Found: C, 72.7; H, 5.6.

Properties

Both   dinitrogendihydridotris(triphenylphosphine)ruthenium(II) and tetrahydridotris(triphenylphosphine)ruthenium(IV) are colorless when pure, but most samples, especially of the nitrogen complex, have a slight tan to light-pink color. Both complexes are stable indefinitely as solids in inert atmospheres; both are decomposed by air. In solution, an atmosphere of the appropriate gas, nitrogen or hydrogen, is required for prolonged stability.

The infrared spectrum of the nitrogen complex (Nujol mull) exhibits a strong sharp band at 2147 cm.$^{-1}$ ($N \equiv N$ stretch) and bands of moderate intensity at 1947 and 1917 cm.$^{-1}$ assignable to ruthenium-hydrogen stretching. The infrared spectrum of the tetrahydride includes a somewhat broad band centered at 1950 cm.$^{-1}$, assigned to ruthenium-hydrogen stretching.

The nitrogen complex is converted rapidly to the tetrahydride by passing a stream of hydrogen through its solutions; similarly, passage of nitrogen through solutions of the tetrahydride generates the nitrogen complex. These reactions also occur in the solid state, though somewhat less readily. Reversible exchange of both species also occurs with ammonia, the ammoniated product being formulated as $RuH_2(NH_3)[(C_6H_5)_3P]_3$. Reaction of the nitrogen complex with triphenylphosphine liberates nitrogen and forms RuH_2-$[(C_6H_5)_3P]_4$; reaction of the tetrahydride with triphenylphosphine liberates hydrogen to form the same dihydride.

References

1. (a) A. Yamamoto, S. Kitazume, and S. Ikeda, *J. Am. Chem. Soc.*, **90**, 1089 (1968).
   (b) T. Ito, S. Kitazume, A. Yamamoto, and S. Ikeda, *ibid.*, **92**, 3011 (1970).
2. W. H. Knoth, *ibid.*, **94**, 104 (1972).
3. R. A. Schunn and E. R. Wonchoba, *Inorganic Syntheses*, **13**, 131 (1972).
4. D. F. Shriver, "The Manipulation of Air-sensitive Compounds," McGraw-Hill Book Company, New York, 1969.

10. PENTAHYDRIDOBIS(TRIMETHYLPHOSPHINE)- IRIDIUM(V)

Submitted by E. KENT BAREFIELD*
Checked by U. KLABUNDE†

A number of multihydride complexes of iridium are known. One series of complexes which contains only phosphine and hydride donors was initially reported to have the stoichiometry (phosphine)$_2$-IrH_3.[1,2] The complex with diethylphenylphosphine[3] and the present example with trimethylphosphine have been shown to be pentahydrido species. These are prepared generally by the reaction of a reducing agent, such as $LiAlH_4$, with $IrX_3(PR_3)_3$ or $[R_3PH]$-$[IrX_4(PR_3)_3]_2$ in which X is halogen and R is alkyl or aryl. With the exception of the triphenylphosphine complex,[4] which is extremely

*Department of Chemistry, University of Illinois, Urbana, Ill. 61801.
†Central Research Department, E. I. du Pont de Nemours & Company, Wilmington, Del. 19898.

insoluble, the other derivatives are very soluble in all organic solvents and their isolation is difficult. The trimethylphosphine complex is unique in that it is quite volatile *in vacuo* and can be sublimed at room temperature.

A. TRIMETHYLPHOSPHONIUM TETRACHLOROBIS-(TRIMETHYLPHOSPHINE)IRIDATE(1—) AND TRICHLOROTRIS(TRIMETHYLPHOSPHINE)IRIDIUM(III)[5]

$$H_3IrCl_6 + (CH_3)_3P \longrightarrow$$

$$[(CH_3)_3PH][IrCl_4\{P(CH_3)_3\}_2] + IrCl_3[P(CH_3)_3]_3 + [(CH_3)_3PH]Cl$$

Procedure

■ **Caution.** *Trimethylphosphine is toxic and spontaneously flammable. It should be transferred by vacuum-line techniques,[6] and the entire reaction should be performed in an efficient hood.*

A solution of 15 g. (30 mmoles) of hexachloroiridic acid* (38.69 % Ir) and 1.5 ml. of concentrated hydrochloric acid in 150 ml. of 2-methoxyethanol is refluxed under nitrogen for about 30 minutes until the solution turns from brown to green. This solution is cooled and then transferred by syringe to a nitrogen-filled 200-ml. Carius tube. The tube is attached to a vacuum manifold, cooled to − 78°, and evacuated. Trimethylphosphine, *ca.* 9 g. (118 mmoles), is condensed into the tube, and the neck is sealed with a torch. The tube is warmed to room temperature, placed in a jacket, and heated in a steam bath for one hour. After cooling, the tube is opened and the contents filtered through a medium frit. The filtrate is evaporated to *ca.* 30 ml. to give pink crystals, which are collected, washed with ethanol and ether successively, and dried under high vacuum. The yield of crude $[(CH_3)_3PH][IrCl_4\{P(CH_3)_3\}_2]$ suitable for use in Sec. B is 7 g. (41 %). The salt may be crystallized from dimethyl sulfoxide solution by addition of ether. *Anal.* Calcd. for $C_9H_{28}IrP_3Cl_4$: C, 19.2; H, 5.01; P, 16.47; Cl, 25.18. Found: C, 19.71; H, 4.60; P, 16.58; Cl, 24.93.

The filtrate from collection of the salt contains a red-yellow oil. Refluxing this oil in 2-methoxyethanol, followed by evaporation,

*Available from Engelhard Industries, 429 Delancy St., Newark, N.J. 07105.

gives a yellow oil which yields yellow crystals of *mer*-IrCl$_3$[P-(CH$_3$)$_3$]$_3$ upon standing. (Identical material can be obtained by refluxing the pink phosphonium salt in 2-methoxyethanol for 25 minutes.) The yellow solid is recrystallized from chloroform. *Anal.* Calcd. for C$_9$H$_{27}$IrP$_3$Cl$_3$: C, 20.65; H, 5.16. Found: C, 20.95; H, 5.25. After refluxing the yellow oil in 2-methoxyethanol and collecting the yellow meridional isomer, the residue remaining is dissolved in ethanol and allowed to stand. White crystals of *fac*-IrCl$_3$[P(CH$_3$)$_3$]$_3$ deposit. *Anal.* Found: C, 20.16; H, 5.12.

Properties

The phosphonium salt and the two isomeric IrCl$_3$[PCH$_3$)$_3$]$_3$ complexes are air-stable in the crystalline state. The ^{1}H n.m.r. spectrum of [(CH$_3$)$_3$PH][IrCl$_4${P(CH$_3$)$_3$}$_2$] in dimethyl sulfoxide-d$_6$ shows a doublet ($J = 15$ Hz.) at τ8.16 and a triplet ($J = 4$ Hz.) at τ8.58 with about twice the intensity of the doublet. The spectrum of this solution is time-dependent. After several days, the methyl region becomes very complex.

The ^{1}H n.m.r. spectrum of the yellow complex in chloroform-d consists of a doublet at τ8.33 ($J = 10$ Hz.) and a triplet at τ8.38 ($J = 4$ Hz.). The infrared spectrum contains bands for $\nu_{\text{Ir—Cl}}$ at 320 and 260 cm.$^{-1}$. These data are consistent with the assignment of the meridional or trans configuration to this compound. The white isomer is insoluble in most organic solvents.

B. PENTAHYDRIDOBIS(TRIMETHYLPHOSPHINE)-IRIDIUM(V)

$$[(CH_3)_3PH][IrCl_4\{P(CH_3)_3\}_2] + LiAlH_4 \longrightarrow$$

$$IrH_5[P(CH_3)_3]_2 + LiAlCl_4 + (CH_3)_3P$$

Procedure

A 250-ml. three-necked flask is fitted with a magnetic stirring bar and a condenser surmounted by an N$_2$ inlet and bubbler system. The flask is charged with 1 g. (1.7 mmoles) of trimethylphosphonium tetrachlorobis(trimethylphosphine)iridate(1 −) [an equivalent amount of trichlorotris(trimethylphosphine)iridium(III) may be used] and 60 ml. of rigorously anhydrous tetrahydrofuran. Lithium tetra-hydroaluminate (0.35 g.) is added in small portions with good stirring. After the addition is complete (5 minutes), the mixture is boiled under reflux for one hour. The mixture is then cooled to room temperature and hydrolyzed with water in tetrahydrofuran (*ca.* 2 ml. of water in 30 ml. of THF).* Filtration to remove solids is carried out under nitrogen, and the filtrate is collected in a single-necked flask. The solvent is evaporated under reduced pressure, leaving a yellow semisolid. The flask is then fitted with a water-cooled cold finger and evacuated. White crystals of pentahydridobis (trimethylphosphine)iridium(V) grow on the cold finger overnight. These are removed under an inert atmosphere to give about 0.45 g. of product (75–85% yield). {The checker obtained 0.54 g. (87%) from the phosphonium salt or 0.46 g. (74%) from *mer*-IrCl$_3$[P-(CH$_3$)$_3$]$_3$. He found it necessary to warm the crude solid to 50° to complete the sublimation.} *Anal.* Calcd. for C$_6$H$_{23}$IrP$_2$: C, 20.62; H, 6.63. Found: C, 21.04; H, 6.66.

Properties

Pentahydridobis(trimethylphosphine)iridium(V) is a white crystal-line solid which melts with decomposition at 70°. It decomposes rapidly in air and slowly in an inert atmosphere, though repurifi-cation through sublimation is easy. The infrared spectrum contains a strong metal-hydrogen stretching absorption at 1920 cm.$^{-1}$. The ^{1}H spectrum in toluene-d$_8$ consists of a pseudo-triplet of intensity 18 at $\tau 8.26$ ($J_{P-CH_3} = 4$ Hz.) for the methyl protons and a triplet of intensity 5.5 at $\tau 19.73$ ($J_{P-IrH} = 14$ Hz.). The ^{31}P spectrum, after selective decoupling of the methyl protons, consists of a sextet at + 192.1 p.p.m. from external trimethyl phosphite.

In benzene the complex slowly decomposes with elimination of hydrogen to give a new species whose structure is uncertain. The pentahydride (or its decomposition products) catalyzes the exchange of benzene hydrogen with gaseous deuterium.[8] Reaction with trimethylphosphine is rapid with loss of hydrogen to yield a

---

*It has been reported that the addition of diethyl ether at this point aids in filtration and in removal of excess water in the evaporation step.[7] The addition of 40–50 ml. of ether may be helpful but is not essential.

mixture containing fac-$[(CH_3)_3P]_3IrH_3$ and probably mer-$[(CH_3)_3$-$P]_3IrH_3$. Olefins are hydrogenated and isomerized by the complex in the absence of hydrogen. In the presence of hydrogen the complex acts as both an isomerization and hydrogenation catalyst.

References

1. L. Malatesta, G. Caglio, and M. Angoletta, *J. Chem. Soc.*, **1965**, 6974.
2. J. Chatt, R. S. Coffey, and B. L. Shaw, *ibid.*, **1965**, 7391.
3. B. E. Mann, C. Master, and B. L. Shaw, *Chem. Commun.*, **1970**, 703.
4. N. Ahmad, J. J. Levision, S. D. Robinson, and M. P. Uttley, *Inorganic Syntheses*, **15**, 45 (1974).
5. J. Chatt, A. E. Field, and B. L. Shaw, *J. Chem. Soc.*, **1963**, 3371.
6. D. F. Shriver, "The Manipulation of Air-sensitive Compounds," McGraw-Hill Book Company, New York, 1969.
7. J. Chatt and R. S. Coffey, *J. Chem. Soc. (A)*, **1969**, 1963.
8. E. K. Barefield, G. W. Parshall, and F. N. Tebbe, *J. Am. Chem. Soc.*, **92**, 5234 (1970).

11. HYDRIDE COMPLEXES OF IRON(II) AND RUTHENIUM(II)

Submitted by W. G. PEET* and D. H. GERLACH*
Checked by D. D. TITUS †

Tetrakis(organophosphorus)metal hydrides (containing one, two, three, or four hydride nuclei) have recently attracted considerable attention in connection with their stereochemical behavior[1] and reaction chemistry.[2] Detailed nuclear magnetic resonance studies and line-shape analyses have substantiated stereochemical non-rigidity in many of these complexes in solution and have yielded mechanistic information about the intramolecular-exchange processes.[3]

Bis[ethylenebis(diphenylphosphine)]dihydridoiron(II) has been prepared by a two-step procedure that involves reduction of iron(II) chloride by ethoxydiethylaluminum.[4] The procedure described be-

*Central Research Department, E. I. du Pont de Nemours & Company, Wilmington, Del. 19898.
† Department of Chemistry, Temple University, Philadelphia, Pa. 19122.

low avoids the use of the alkyl aluminum compound and gives the iron hydride in one step.[5] This procedure and that for dihydrido-tetrakis(triethyl phosphite)ruthenium(II) use sodium tetrahydro-borate as the reducing agent as in the previously described synthesis of tetrakis(diethyl phenylphosphonite)dihydridoiron(II).[6]

A. BIS[ETHYLENEBIS(DIPHENYLPHOSPHINE)]-DIHYDRIDOIRON(II)

$$FeCl_2 + 2(C_6H_5)_2PCH_2CH_2P(C_6H_5)_2 \xrightarrow{NaBH_4}$$

$$FeH_2[(C_6H_5)_2PCH_2CH_2P(C_6H_5)_2]_2$$

Procedure

Highly reactive anhydrous iron(II) chloride is essential in this preparation. Iron(II) chloride tetrahydrate may be dehydrated under vacuum or by azeotropic distillation. In either case the highly hygro-scopic $FeCl_2$ must be protected from moist air before use. In vacuum drying, the $FeCl_2(H_2O)_4$ is thoroughly ground under nitrogen and is dried at $110°$ for 3–4 hours at *ca.* 0.1 torr in an Abderhalden pistol containing Drierite.

In the azeotropic-distillation method, 50 g. of iron(II) chloride tetrahydrate is finely ground in an inert atmosphere and is placed in a round-bottomed flask containing 250 ml. of benzene. The water is azeotropically distilled from the iron(II) chloride under nitrogen for 48 hours. The resulting tan solid is separated by filtration and vacuum-dried at $25°$ for 6 hours.

■ **Caution.** *The dihydridoiron complex is very oxygen-sensitive; so the reduction and isolation steps described below must be carried out in a dry-box or by use of Schlenk-tube techniques.*[7]

A 500-ml., single-necked, round-bottomed flask equipped with a reflux condenser and magnetic stirring bar is charged with 250 ml. of anhydrous tetrahydrofuran, 20 g. (0.05 mole) of ethylenebis-[diphenylphosphine](1, 2-bis(diphenylphosphino)ethane),* and 2.52 g. (0.02 mole) of anhydrous iron(II) chloride. The tan mixture is stirred for 10 minutes, and 4.0 g. (0.11 mole) of sodium tetrahydro-borate* is added. The purple-red reaction mixture is heated to reflux,

*Available from Alfa Inorganics, P.O. Box 159, Beverly, Mass. 01915.

and an additional 2.0 g. (0.055 mole) of sodium tetrahydroborate is added, followed by slow addition of 20 ml. of absolute ethanol. After gas evolution ceases, the chrome-yellow solution is refluxed for 10 minutes and filtered. The volume of the filtrate is reduced by 80% by evaporation under reduced pressure, and the resulting yellow crystals are collected and dried under vacuum. Recrystallization by dissolving the product in 500 ml. of toluene, filtering, and reducing the volume of the filtrate by 60% yields pure FeH_2-$[(C_6H_5)_2PCH_2CH_2P(C_6H_5)_2]_2$ as a toluene solvate. After drying *in vacuo* at 50° for 8 hours, 11.5 g. (56%) of the stoichiometric solvate $H_2Fe[(C_6H_5)_2PCH_2CH_2P(C_6H_5)_2]_2 \cdot 2C_7H_8$ is obtained. *Anal.* Calcd. for $C_{66}H_{64}P_4Fe$: C, 76.3; H, 6.40; P, 11.92; Fe, 5.37. Found: C, 76.0; H, 6.53; P, 11.93; Fe, 5.37.

Properties

The product $H_2Fe[(C_6H_5)_2PCH_2CH_2P(C_6H_5)_2]_2 \cdot 2C_7H_8$ is a light-yellow crystalline solid which is unstable toward atmospheric oxygen. The solid can be exposed to air for a few minutes without visible decomposition. Hydrocarbon solutions are significantly less stable, decomposing visibly upon even brief exposure to air. The compound is moderately soluble in tetrahydrofuran and dichloromethane, decomposing slowly in the latter solvent. It is slightly soluble in toluene, benzene, and ethanol and virtually insoluble in petroleum ether and *n*-heptane. Its infrared spectrum contains Fe-H stretching absorption at 1840 cm.$^{-1}$.

B. TETRAKIS(TRIETHYL PHOSPHITE)-DIHYDRIDORUTHENIUM(II)

$$RuCl_3 \cdot 3H_2O + 4P(OC_2H_5)_3 \xrightarrow{\text{NaBH}_4} RuCl_2[P(OC_2H_5)_3]_4$$

$$RuCl_2[P(OC_2H_5)_3]_4 + P(OC_2H_5)_3 \xrightarrow{\text{NaBH}_4} RuH_2[P(OC_2H_5)_3]_4$$

Procedure

■ **Caution.** *The product is moderately air-sensitive; so all operations should be carried out in an inert atmosphere.*

A 200-ml., single-necked, round-bottomed flask equipped with a magnetic stirring bar is charged with 25 ml. (excess) of triethyl phosphite and 2.6 g.(0.01 mole) of "ruthenium chloride trihydrate."* The mixture is stirred for 15 minutes, during which time it turns violet and becomes hot. After addition of 1 g. (0.026 mole) of sodium tetrahydroborate† and stirring for 5 minutes, the yellow solution is filtered. The volume of the filtrate is reduced by 60% by evaporation under reduced pressure, and the resulting yellow crystals are collected and dried *in vacuo* at room temperature for 20 hours. The crystals are purified by dissolving in 500 ml. of *n*-hexane, filtering, and reducing the volume of the filtrate by 70%. The yellow crystals which form after cooling at 0° for several hours are collected and dried *in vacuo*. The yield of $RuCl_2[P(OC_2H_5)_3]_4$ is 5.0 g.(63%). *Anal.* Calcd. for $C_{24}H_{60}O_{12}P_4Cl_2Ru$: C, 34.45; H, 7.23; P, 14.81; Cl, 8.47; Ru, 12.08. Found: C, 34.29; H, 7.33; P, 14.73; Cl, 8.69; Ru, 11.00.

A 200-ml. Erlenmeyer flask containing a magnetic stirring bar is charged with 5 g. (0.006 mole) of tetrakis(triethyl phosphite) dichlororuthenium(II), 2 ml. (0.013 mole) of triethyl phosphite, and 75 ml. of ethanol. This solution, after addition of 3 g.(0.079 mole) of sodium tetrahydroborate, is refluxed gently for 30 minutes and filtered. The solvent is evaporated from the filtrate, and the gummy yellow residue is extracted with four 50-ml. portions of petroleum ether. The petroleum ether is evaporated *in vacuo*, and the resulting solid is recrystallized once by dissolving in *ca.* 50 ml. of absolute ethanol and cooling to − 40° for 2 days. The white crystalline product is collected and dried *in vacuo* for 3 hours. The yield is 2.6 g. (51%). *Anal.* Calcd. for $C_{24}H_{62}O_{12}P_4Ru$: C, 37.55; H, 8.14; O, 25.91; P, 16.14. Found: C, 37.96, 37.77; H, 8.18, 8.26; O, 24.83, 24.82; P, 16.73.

Properties

Tetrakis(triethyl phosphite)dihydridoruthenium(II) is a white crystalline solid which is slightly unstable toward atmospheric oxygen. The solid can be exposed to air for several minutes without visible decomposition. Solutions are more unstable and decompose visibly

*Engelhard Industries, 429 Delancy St., Newark, N.J. 07105.

†Alfa Inorganics, P.O. Box 159, Beverly, Mass. 01915. With less reactive Na[BH₄], *ca.* 30 minutes is required for attainment of the yellow color.

upon exposure to air. The compound is soluble in common hydrocarbon solvents, ethanol, and tetrahydrofuran. Its infrared spectrum shows v_{Ru-H} at 1880 cm.$^{-1}$.

References

1. J. P. Jesson, in "Transition Metal Hydrides," E. L. Muetterties (ed.), vol. 1, Marcel Dekker, Inc., New York, 1971.
2. R. A. Schunn, *ibid.*, Chap. 5; C. A. Tolman, *ibid.*, Chap. 6.
3. P. Meakin, E. L. Muetterties, F. N. Tebbe, and J. P. Jesson, *J. Am. Chem. Soc.*, **93**, 4701 (1971).
4. G. Hata, H. Kondo, and A. Miyake, *ibid.*, **90**, 2278 (1968).
5. D. H. Gerlach, W. G. Peet, and E. L. Muetterties, *ibid.*, **94**, 4545 (1972).
6. D. D. Titus, A. A. Orio, and H. B. Gray, *Inorganic Syntheses*, **13**, 117 (1971).
7. J. J. Eisch and R. B. King (eds.), "Organometallic Syntheses," vol. 1, Academic Press, Inc., New York, 1965; D. F. Shriver, "The Manipulation of Air-sensitive Compounds," McGraw-Hill Book Company, New York, 1969.

12. TETRAHYDRIDOTETRAKIS-(METHYLDIPHENYLPHOSPHINE)MOLYBDENUM(IV)

$$[MoCl_4\{P(CH_3)(C_6H_5)_2\}_2] + P(CH_3)(C_6H_5)_2 \xrightarrow[C_2H_5OH]{NaBH_4}$$

$$[MoH_4\{P(CH_3)(C_6H_5)_2\}_4]$$

Submitted by FILIPPO PENNELLA*
Checked by W. G. PEET†

The method for the synthesis of the title compound is applicable to the synthesis of analogous tertiary phosphine and diphosphine tetrahydride complexes of molybdenum: thus $[MoH_4\{P(C_2H_5)(C_6H_5)_2\}_4]$ and $[MoH_4\{(C_6H_5)_2PCH_2CH_2P(C_6H_5)_2\}_2]$ have been prepared similarly.[1] The tetrahydrides are obtained by the reaction of $[MoCl_4(PR_3)_2]$, where $PR_3 = P(CH_3)(C_6H_5)_2$ or $P(C_2H_5)(C_6H_5)_2$, or of $[MoCl_4\{(C_6H_5)_2PCH_2CH_2P(C_6H_5)_2\}_2]$ with $NaBH_4$ in ethanol in the presence of the appropriate free ligand.

*Phillips Petroleum Company, Research and Development Department, Bartlesville, Okla. 74004.

†Central Research Department, E. I. du Pont de Nemours & Company, Wilmington, Del. 19898.

The tetrachloro complexes are prepared by the reaction of $[MoCl_4-(NCC_2H_5)_2]$ with the phosphine in dichloromethane, following a procedure similar to that reported for $[MoCl_4\{P(C_6H_5)_3\}_2]$.[2]

A. TETRACHLOROBIS(METHYLDIPHENYLPHOSPHINE)-MOLYBDENUM(IV)

$[MoCl_4(NCC_2H_5)_2] + P(CH_3)(C_6H_5)_2 \longrightarrow$

$$[MoCl_4\{P(CH_3)(C_6H_5)_2\}_2]$$

Procedure

The reaction is carried out in an inert atmosphere using rigorously anhydrous solvent. Tetrachlorobis(propionitrile)molybdenum(IV)[3] (1.7 g., 5 mmoles) is added to 20 ml. of dichloromethane containing 2.2 g. (11 mmoles) of methyldiphenylphosphine. The solid dissolves immediately, and within 1 minute, dark-red crystals begin to form. After 20 minutes the product is collected by filtration, washed twice with 8 ml. of dichloromethane, and dried under vacuum. The yield is approximately 2.5 g. (77%). The product, when pulverized, is dark orange. *Anal.* Calcd. for $C_{26}H_{26}P_2Cl_4Mo$: C, 48.9; H, 4.11. Found: C, 48.8; H, 4.10.

B. TETRAHYDRIDOTETRAKIS-(METHYLDIPHENYLPHOSPHINE)MOLYBDENUM(IV)

$[MoCl_4\{P(CH_3)(C_6H_5)_2\}_2] + P(CH_3)(C_6H_5)_2 \xrightarrow[C_2H_5OH]{NaBH_4}$

$$[MoH_4\{P(CH_3)(C_6H_5)_2\}_4]$$

Procedure

The reaction is carried out in an atmosphere of scrupulously dry, air-free argon or helium.* Tetrachlorobis(methyldiphenylphosphine)molybdenum(IV) (1.6 g., 2.5 mmoles) is added slowly with stirring to 30 ml. of ethanol containing 1.2 g. (6 mmoles) of methyldiphenylphosphine and 0.8 g. (21 mmoles) of sodium tetra-

---

*The checker found the reaction to be N_2-sensitive. None of the desired product was obtained under an N_2 atmosphere, but a 55% crude yield was obtained under argon.

hydroborate ($NaBH_4$). After the addition is completed (10 to 20 minutes), stirring is continued for one hour. The mixture is filtered by suction, and the filtrate is discarded. Twice the solid is placed in 20 ml. of methanol, stirred for 10 minutes, and filtered. The yield of the crude product is 1.7–1.8 g. (75–80%). The product is dissolved in 25 ml. of benzene at room temperature, the solution is filtered, and methanol (*ca.* 30 ml.) is added slowly to the filtrate to precipitate the product. The product is collected by filtration, washed with MeOH, and dried in vacuum. Yield is 1.0 g. (55–60%; over-all 45%), m.p. 166–169°. *Anal.* Calcd. for $C_{52}H_{56}P_4Mo$: C, 69.3; H, 6.27. Found: C, 69.2; H, 6.4.

Properties

Tetrahydridotetrakis(methyldiphenylphosphine)molybdenum(IV) is a yellow solid, moderately soluble in benzene, toluene, and tetrahydrofuran. The infrared spectrum (Nujol mull) shows bands in the Mo-H stretching region at 1800 and 1714 cm.$^{-1}$ and in the bending region at 640 and 775 cm.$^{-1}$. In the proton n.m.r. spectrum in deuterobenzene solution the hydridic proton signal appears as a 1:4:6:4:1 quintet centered at $\tau 12.25$, J(P—H)33 Hz. The temperature dependence of the ^1H and ^{31}P spectra has been interpreted[4] to indicate a stereochemically nonrigid structure with interpenetrating H_4 and P_4 tetrahedra.

References

1. F. Pennella, *Chem. Commun.*, **1971**, 158.
2. E. A. Allen, K. Feenan, and G. W. Fowles, *J. Chem. Soc.*, **1965**, 1636.
3. E. A. Allen, B. J. Brisolon, and G. W. Fowles, *ibid.*, **1964**, 4531.
4. J. P. Jesson, E. L. Muetterties, and P. Meakin, *J. Am. Chem. Soc.*, **93**, 5261 (1971).

Chapter Three

TRIPHENYLPHOSPHINE COMPLEXES OF TRANSITION METALS

13. COMPLEXES OF RUTHENIUM, OSMIUM, RHODIUM, AND IRIDIUM CONTAINING HYDRIDE CARBONYL, OR NITROSYL LIGANDS

Submitted by N. AHMAD,* J. J. LEVISON,* S. D. ROBINSON,* and M. F. UTTLEY*
Checked by E. R. WONCHOBA† and G. W. PARSHALL†

The increasing utility of triphenylphosphine derivatives of the platinum-group metals as homogeneous catalysts and as starting materials in preparative chemistry highlights the desirability of convenient small-scale syntheses for these compounds. Synthetic procedures in current use are frequently based on early methods involving heterogeneous reaction conditions and requiring use of multistage syntheses, prolonged reaction times, or dangerous reagents. The procedures described below are developed from a technique of synthesis devised by the present authors,[1,2] and permit the rapid synthesis of a wide range of triphenylphosphine derivatives of the platinum-group metals under homogeneous reaction conditions. The essentially homogeneous nature of each reaction solution,

*Department of Chemistry, King's College, Strand, London, WC2R 2LS, England.
Present addresses: Department of Chemistry, Aligarh Muslim University, Aligarh, U.P. India (N.A.): International Nickel Limited, Bashley Road, London, N.W. 10, England (J.J.L.).
†Central Research Department, E. I. du Pont de Nemours & Company, Wilmington, Del. 19898.

which is maintained until precipitation or crystallization of the required product occurs, renders these syntheses rapid, efficient, and highly selective. Thus variations of the basic technique involving changes in reaction temperature and solvent, concentration of ligands, and nature of added reagents provide routes to a range of pure triphenylphosphine derivatives. Moreover, the low solubility of most triphenylphosphine complexes in the alcoholic solvents employed in these reactions ensures the rapid precipitation or crystallization of the required product in good yield. The speed and selectivity are illustrated by the experimental details and the range of syntheses given. The efficiency is revealed in the high yields and good analytical data, the latter frequently obtained without recourse to purification procedures. Because of solubility considerations, these techniques are specific to the preparation of triphenylphosphine complexes.

A discussion of the basic technique, apparatus, and reagents employed together with details concerning methods of purification and characterization of products is given in the following section. The success of these syntheses is dependent on strict adherence to the procedures described.

General Technique

Platinum-group metal salts, $RuCl_3 \cdot 3H_2O$ (39% Ru),* $Na_2OsCl_6 \cdot 6H_2O$ (34% Os),† $RhCl_3 \cdot 3H_2O$ (39% Rh), and $Na_2IrCl_6 \cdot 6H_2O$ (34% Ir) are supplied by Johnson Matthey Limited and Engelhard Industries. Sodium tetrahydroborate or alcoholic potassium hydroxide, 40% w/v aqueous formaldehyde solution, and N-methyl-N-nitroso-p-toluenesulfonamide serve as sources of hydride, carbonyl,[3] and nitrosyl ligands, respectively. The alcoholic reaction media are ethanol and 2-methoxyethanol, which are used without further purification. All reactions are performed in a 250-ml. conical (Erlenmeyer) reaction flask bearing an adapter fitted with a 30-cm. con-

---

*Commercial hydrated ruthenium trichloride has roughly this composition but contains a considerable number of distinct chemical species.

†The readily available $(NH_4)_2OsCl_6$ is easily converted to the more soluble sodium salt by passage through a sodium-loaded sulfonic-acid-type ion-exchange resin (such as Rexyn 101) in hot aqueous solution.

denser, a nitrogen inlet, and a stoppered port for the rapid introduction of reagent solutions. The flask is situated on a hot plate equipped with a magnetic stirrer to provide efficient heating and stirring of the reaction solutions.

The basic technique employed is common to all the syntheses described in this chapter and involves the *rapid, successive* addition of alcoholic solutions of the appropriate platinum-metal salt and other reagents to a vigorously stirred, boiling, alcoholic solution of triphenylphosphine, which is subsequently heated under reflux until precipitation of the required product commences or until the reaction is complete. In the latter instances the required products crystallize cleanly from the reaction solution on cooling. The success of these syntheses is critically dependent upon the maintenance of essentially homogeneous reaction conditions until the reaction sequence is complete. This in turn requires the addition of freshly prepared reagent solutions rapidly and successively in the specified order to the vigorously stirred, boiling reaction solution. Failure to observe this precaution leads to precipitation of insoluble intermediates, which may fail to react further and hence contaminate the reaction product.

The products are generally washed successively with ethanol (2 × 10 ml.), water (2 × 10 ml.), ethanol (2 × 10 ml.), and *n*-hexane (10 ml.) and dried *in vacuo*. Where necessary, further purification of products is achieved by recrystallization using solvents recommended in the literature (refer to individual syntheses for details). Products are characterized by analytical, spectroscopic, and melting-point data. The reported infrared spectra were recorded as Nujol mulls on a Perkin-Elmer 457-grating spectrometer. N.m.r. spectra were recorded on a Varian HA100 spectrometer with tetramethylsilane used as an internal standard. Although the complexes are all apparently air-stable in the crystalline state, all show appreciable *lowering* of melting point when recrystallized repeatedly from organic solvents in the presence of air. Recorded melting points correspond to the highest observed for a particular complex, and all are accompanied by decomposition. All complexes discussed are insoluble in alcohol, ether, or saturated hydrocarbon solvents; solubilities in benzene, chloroform, and dichloromethane are listed for specific complexes.

A. CARBONYLCHLOROHYDRIDOTRIS(TRIPHENYL-PHOSPHINE) RUTHENIUM(II)

$$RuCl_3 + 3(C_6H_5)_3P \xrightarrow{HCHO} RuHCl(CO)[P(C_6H_5)_3]_3$$

Carbonylchlorohydridotris(triphenylphosphine)ruthenium(II) was originally prepared by prolonged reaction of ruthenium trichloride and triphenylphosphine in basic alcoholic media.[4,5] It is more conveniently obtained from the homogeneous reaction of triphenylphosphine, ruthenium trichloride, and aqueous formaldehyde in boiling 2-methoxyethanol.[1]

Procedure

Solutions of 0.26 g. (1.0 mmole) of hydrated ruthenium trichloride in 20 ml. of 2-methoxyethanol and aqueous formaldehyde (20 ml., 40% w/v solution) are added rapidly and successively to a vigorously stirred, boiling solution of 1.58 g. (6 mmoles) of triphenylphosphine in 60 ml. of 2-methoxyethanol. The mixture is heated under reflux for 10 minutes and allowed to cool. The precipitate which forms is separated and washed successively with ethanol, water, ethanol, and *n*-hexane and dried *in vacuo*. Yield: 0.89 g. (93% based on $RuCl_3 \cdot 3H_2O$). *Anal*. Calcd. for $C_{55}H_{46}ClOP_3Ru$: C, 69.36; H, 4.87; Cl, 3.72; P, 9.76. Found: C, 69.3; H, 4.94; Cl, 3.36; P, 9.72.

Properties

Carbonyldihydridotris(triphenylphosphine)ruthenium(II) forms cream-white microcrystals which melt at 209–210° in air and at 235–237° in a capillary sealed under nitrogen. The infrared spectrum shows bands at 2020 (m), 1922 (vs), and 1903 (sh) cm.$^{-1}$ attributed to $v(CO)$ and $v(RuH)$. The complex is very sparingly soluble in benzene and slightly soluble in chloroform.

B. CARBONYLDIHYDRIDOTRIS(TRIPHENYLPHOSPHINE)-RUTHENIUM(II) (WHITE ISOMER)

$$RuCl_3 + 3(C_6H_5)_3 P \xrightarrow[KOH]{HCHO} RuH_2(CO)[P(C_6H_5)_3]_3$$

Carbonyldihydridotris(triphenylphosphine)ruthenium(II) was originally prepared[6] by boiling $RuCl_2[P(C_6H_5)_3]_3$ with sodium tetrahydroborate in benzene-ethanol or benzene-methanol (3:5) under hydrogen for one hour, then cooling the solution in a stream of hydrogen when white platelets of the required product crystallize during 15 hours. The procedure described below is a convenient single-step synthesis.[1]

Procedure

Solutions of 0.52 g. (2 mmoles) of hydrated ruthenium trichloride in 20 ml. of ethanol, aqueous formaldehyde (20 ml., 40% w/v solution), and 0.6 g. of potassium hydroxide in 20 ml. of ethanol are added quickly and successively to a boiling solution of 3.14 g. (12 mmoles) of triphenylphosphine in 140 ml. of ethanol. The solution is heated under reflux for 15 minutes and then cooled. The resultant gray precipitate is separated, washed successively with ethanol, water, ethanol, and *n*-hexane, and then dried *in vacuo*. Yield 1.3 g. (70% based on $RuCl_3 \cdot 3H_2O$).

The crude precipitate is dissolved in the minimum volume of warm benzene, and the solution is filtered and passed through a 9-in. × 1-in. column of activated neutral alumina. The benzene eluate is diluted with methanol, concentrated under vacuum at 25°, and then set aside to crystallize at 0°. Crystals of pure carbonyldihydridotris-(triphenylphosphine)ruthenium(II) are filtered, washed with *n*-hexane, and dried *in vacuo*. *Anal.* Calcd. for $C_{55}H_{47}OP_3Ru$: C, 71.96; H, 5.16; P, 10.12. Found: C, 72.39, 71.8; H, 5.13, 5.2; P, 10.43, 9.7.

Recrystallization from dichloromethane-methanol affords a dichloromethane adduct, $RuH_2(CO)[P(C_6H_5)_3]_3 \cdot CH_2Cl_2$. *Anal.* Calcd. for $C_{56}H_{49}Cl_2OP_3Ru$: C, 67.0; H, 4.9; Cl, 7.05. Found: C, 67.0; H, 4.9; Cl, 6.9.

Properties

Carbonyldihydridotris(triphenylphosphine)ruthenium(II) forms white flaky microcrystals which melt at 160–162° in air and at 213–215° in a capillary sealed under nitrogen. The infrared spectrum shows bands at 1960 (m) and 1898 (m) cm.$^{-1}$, attributed to ν(RuH), and at 1940 (vs) cm.$^{-1}$, attributed to ν(CO). The high-field n.m.r. spectrum (benzene solution) comprises a complex first-order pattern of 24 lines indicative of rigid stereochemistry (I).

H_A	$\tau 16.47$	$J(H_A P_D)$ 15.5 Hz.
		$H(H_A P_C)$ 30.0 Hz.
		$J(H_A H_B)$ 6.0 Hz.
H_B	$\tau 18.33$	$J(H_B P_D)$ 73.5 Hz.
		$J(H_B P_C)$ 28.5 Hz.
		$J(H_B H_A)$ 6.0 Hz.

The structure:

$$\begin{array}{c} H_A \quad P_C \\ \backslash \quad / \\ P_D \!-\! Ru \!-\! H_B \\ / \quad \backslash \\ P_C \quad CO \end{array}$$

I

The complex is moderately soluble in benzene, chloroform, and dichloromethane.

C. TRICARBONYLBIS(TRIPHENYLPHOSPHINE)-RUTHENIUM(0)

$$RuCl_3 + 2(C_6H_5)_3P \xrightarrow{\text{KOH, HCHO}} Ru(CO)_3[P(C_6H_5)_3]_2$$

Tricarbonylbis(triphenylphosphine)ruthenium(0) was first prepared by reducing $RuCl_2(CO)_2[P(C_6H_5)_3]_2$ with zinc dust in dimethylformamide at 100° under a carbon monoxide pressure of 4 atmospheres for 24 hours.[7] It has subsequently been prepared by treating $[Ru(CO)_3\{P(C_6H_5)_3\}]_3$ with excess triphenylphosphine in methyl ethyl ketone solution at 130–140° under an atmosphere of carbon monoxide,[8] and by an autoclave reaction of $Ru(CO)_5$ with triphenylphosphine in tetrahydrofuran at 130° under a nitrogen atmosphere.[9] Passage of carbon monoxide through a methanolic suspension of $RuH(OCOCH_3)[P(C_6H_5)_3]_3$ for *ca.* 12 hours also affords tricarbonylbis(triphenylphosphine)ruthenium(0).[10]

Procedure

Solutions of 0.39 g. (1.5 mmoles) of hydrated ruthenium trichloride in 30 ml. of cool 2-methoxyethanol, hot aqueous formaldehyde (30 ml., 40% w/v), and 0.6 g. of potassium hydroxide in 30 ml. of hot 2-methoxyethanol are added rapidly and successively to a well-stirred, boiling solution of 2.37 g. (9 mmoles) of triphenylphosphine in 90 ml. of 2-methoxyethanol. The reaction solution is stirred and heated under reflux for 1 hour and then cooled to room temperature. The resultant yellow microcrystalline precipitate is washed with ethanol, water, ethanol, and *n*-hexane and dried *in vacuo.* Yield 0.95 g. (89% based on $RuCl_3 \cdot 3H_2O$). *Anal.* Calcd. for

$C_{39}H_{30}O_3P_2Ru$: C, 66.0; H, 4.26; P, 8.73. Found: C, 66.30, 66.0; H, ^{4} 48, 4.2; P, 8.87.

Properties

Tricarbonylbis(triphenylphosphine)ruthenium(0) forms pale yellow microcrystals which melt at 170–173° in air and at 262–266° in a capillary sealed under nitrogen. The infrared spectrum shows a single very strong band at 1900 cm.$^{-1}$ attributable to $v(CO)$. The complex is moderately soluble in benzene and dichloromethane.

D. TRICHLORONITROSYLBIS(TRIPHENYLPHOSPHINE)-RUTHENIUM(II)

$$RuCl_3 + 2(C_6H_5)_3P + p\text{-}TolSO_2N(NO)(CH_3) \longrightarrow$$
$$RuCl_3(NO)[P(C_6H_5)_3]_2$$

Trichloronitrosylbis(triphenylphosphine)ruthenium(II) was originally prepared by heating a mixture of $[RuCl_3(NO)]_n$ and triphenylphosphine in ethanol for 5–10 minutes.[11] It has subsequently been prepared by treating the complexes $RuCl(CO)(NO)[P(C_6H_5)_3]_2$[12] or $RuCl(NO)[P(C_6H_5)_3]_2$[13] with chlorine. The procedure described below is based on ruthenium trichloride and eliminates the need to use preformed ruthenium nitrosyl derivatives.

Procedure

Solutions of 0.39 g. (1.5 mmoles) of hydrated ruthenium trichloride in 30 ml. of ethanol and of 0.63 g. of *N*-methyl-*N*-nitroso-*p*-toluenesulfonamide in 30 ml. of ethanol are added in quick succession to a vigorously stirred, boiling solution of 2.37 g. (9 mmoles) of triphenylphosphine in 90 ml. of ethanol. The mixture is refluxed for 5 minutes and then cooled to room temperature. The green platelets deposited are filtered, washed with ethanol, water, ethanol, and *n*-hexane and are dried *in vacuo*. The green, impure complex is extracted with boiling dichloromethane (5 × 100 ml.). The extract is filtered, evaporated to small volume, and then allowed to crystallize at 0°. The crystals are washed with *n*-hexane. Yield is 0.92 g. (80% based on $RuCl_3 \cdot 3H_2O$). *Anal.* Calcd. for $C_{36}H_{30}Cl_3NOP_2Ru$:

C, 56.74; H, 3.97; Cl, 13.96; N, 1.84; P, 8.13. Found: C, 56.2; H, 3.98; Cl, 14.5; N, 1.81; P, 8.45.

Properties

Trichloronitrosylbis(triphenylphosphine)ruthenium(II) forms yellow-orange microcrystals which melt at 215–230° in air and at 290° in a capillary tube under nitrogen. The infrared spectrum shows a single band at 1873 (vs) cm.$^{-1}$ attributed to $v(NO)$. The impure green material shows a second infrared band at 1744 (w) cm.$^{-1}$. The complex is moderately soluble in benzene and dichloromethane.

E. DINITROSYLBIS(TRIPHENYLPHOSPHINE)RUTHENIUM

$$RuCl_3 + 2(C_6H_5)_3P + p\text{-TolSO}_2N(NO)CH_3) + (C_2H_5)_3N \longrightarrow$$
$$Ru(NO)_2[P(C_6H_5)_3]_2$$

Dinitrosylbis(triphenylphosphine)ruthenium was first prepared by the present authors[1,14] using a synthesis similar to the one described below. It has subsequently been obtained by reacting preformed dihydridotris(triphenylphosphine)ruthenium(II) with nitric oxide,[15] and by reaction of $RuCl_2(CO)_2[P(C_6H_5)_3]_2$ with sodium nitrite in dimethylformamide.[16]

Procedure

A solution of 0.26 g. (1 mmole) of hydrated ruthenium trichloride in 20 ml. of ethanol is added to a stirred solution of 1.56 g. (6.0 mmoles) of triphenylphosphine in 60 ml. of boiling ethanol. Triethylamine (*ca.* 4 ml.) is then added dropwise until a deep-purple color develops. *N*-Methyl-*N*-nitroso-*p*-toluenesulfonamide (0.4 g.) in 20 ml. of ethanol and 6 ml. of triethylamine is added rapidly. The mixture is heated under reflux for 5 minutes, then allowed to cool. The gray precipitate is filtered, washed successively with ethanol, water, ethanol, and *n*-hexane and dried *in vacuo* to give the required complex. Yield is 0.55 g. (82% based on $RuCl_3 \cdot 3H_2O$). *Anal.* Calcd. for $C_{36}H_{30}N_2O_2Ru$: C, 63.06; H, 4.41; N, 4.09; P, 9.03. Found C, 62.17; H, 4.13; N, 4.07; P, 8.26. The complex may be recrystallized from dichloromethane-methanol.

Properties

Dinitrosylbis(triphenylphosphine)ruthenium forms orange-red microneedles which melt at 144–145° in air, and at 185–186° under nitrogen. The infrared spectrum shows bands at 1665 and 1619 cm.$^{-1}$. The complex is soluble in dichloromethane, chloroform, and benzene. $Ru(NO)_2[P(C_6H_5)_3]_2$ reacts with two molecules of oxygen to afford $Ru(O_2)(NO_3)(NO)[P(C_6H_5)_3]_2$.[16]

F. CARBONYLCHLOROHYDRIDOTRIS(TRIPHENYL-PHOSPHINE)OSMIUM(II)

$$Na_2OsCl_6 \cdot 6H_2O + 3(C_6H_5)_3P \xrightarrow{HCHO} OsHCl(CO)[P(C_6H_5)_3]_3$$

Carbonylchlorohydridotris(triphenylphosphine)osmium(II) was originally obtained by reacting ammonium hexachloroosmate(IV) and triphenylphosphine in a high-boiling alcoholic solvent[17] and subsequently correctly formulated as a hydridocarbonyl complex.[4,5,18] Optimum conditions reported for the synthesis involved a reaction time of 4.5 hours with temperature up to 190° and employed 2-(2-methoxyethoxy)ethanol–water mixture as solvent.[18]

Procedure

Solutions of 0.67 g. (1.2 mmoles) of sodium hexachloroosmate(IV) 6-hydrate in 35 ml. of 2-methoxyethanol and aqueous formaldehyde (15 ml., 40% w/v solution) are added in rapid succession to a vigorously stirred, boiling solution of 1.97 g. (7.5 mmoles) of triphenylphosphine in 75 ml. of 2-methoxyethanol. The reaction mixture is heated under reflux for 30 minutes, then allowed to cool to room temperature. During heating the color of the reaction mixture changes from brown-red to yellow. The off-white solid which precipitates is filtered, washed with ethanol, water, ethanol, and *n*-hexane, and dried *in vacuo*. Yield is 1.2 g. (95%). *Anal.* Calcd. for $C_{55}H_{46}ClOP_3Os$: C, 63.42; H, 4.45; P, 8.92. Found: C, 63.54; H, 4.7; P, 8.5.

Properties

Carbonylchlorohydridotris(triphenylphosphine)osmium(II) forms white microcrystals which melt with decomposition in air at 179–183° and at 289–290° in a capillary sealed under nitrogen. The infrared spectrum shows bands at 2099 (m) cm.$^{-1}$ attributed to v(OsH) and at 1906 (vs) and 1891 (s) cm.$^{-1}$ attributed to v(CO). The high-field n.m.r. pattern (benzene solution) comprises a doublet of triplets indicative of the stereochemistry II.

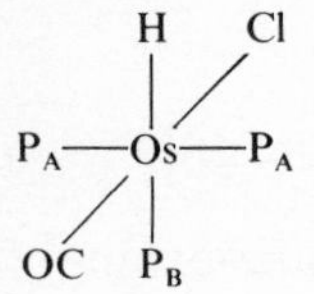

$$P_A \!-\! Os \!-\! P_A$$

H $\tau 16.27$ $J(HP_A)$ 24.5 Hz.

$J(HP_B)$ 87.0 Hz.

II

The complex is moderately soluble in benzene, chloroform, and dichloromethane.

G. CARBONYLDIHYDRIDOTRIS(TRIPHENYLPHOSPHINE)-OSMIUM(II)

$$Na_2OsCl_6 \cdot 6H_2O + 3(C_6H_5)_3P \xrightarrow{KOH, HCHO} OsH_2(CO)\left[P(C_6H_5)_3\right]_3$$

Carbonyldihydridotris(triphenylphosphine)osmium(II) was first prepared by treating preformed $OsHCl(CO)[P(C_6H_5)_3]_3$ with sodium hydroxide in boiling 2-methoxyethanol.[19]

Procedure

Solutions of 0.84 g. (1.5 mmoles) of sodium hexachloroosmate(IV) 6-hydrate in 30 ml. of 2-methoxyethanol, aqueous formaldehyde (12 ml., 40% w/v solution), and 0.84 g. of potassium hydroxide in 15 ml. of 2-methoxyethanol are added rapidly and successively to a vigorously stirred, boiling solution of 4.72 g. (18 mmoles) of triphenylphosphine in 75 ml. of 2-methoxyethanol. The mixture is heated under reflux for 30 minutes, during which period the color changes to pale yellow and a white precipitate separates. The mixture is allowed to cool to room temperature. The off-white precipitate is filtered, washed with ethanol, water, ethanol, and n-hexane, and dried *in vacuo*. Yield is 1.14 g. (73%). *Anal.* Calcd. for $C_{55}H_{47}OOsP_3$: C, 65.60; H, 4.70; P, 9.23. Found: C, 65.83; H, 4.64; P, 8.46.

Properties

Carbonyldihydridotris(triphenylphosphine)osmium(II)   forms white microcrystals which melt at 202–204° in air and at 245–246° in a capillary sealed under nitrogen. The infrared spectrum shows bands at 2043 (m) and 1846 (m) cm.$^{-1}$ attributed to $v(OsH)$, and at 1947 (s) cm.$^{-1}$, attributable to $v(CO)$. The high-field n.m.r. pattern (benzene solution) comprises a complex order pattern of 24 lines.

H_A	τ17.34	$J(H_AP_D)$	16.5 Hz.
		$J(H_AP_C)$	28.5 Hz.
		$J(H_AH_B)$	4.7 Hz.
H_B	τ18.91	$J(H_BP_D)$	69.5 Hz.
		$J(H_BP_C)$	28.5 Hz.
		$J(H_BH_A)$	4.7 Hz.

$$
\begin{array}{c}
H_A \quad P_C \\
\backslash \; / \\
P_D\text{---}Os\text{---}H_B \\
/ \; | \\
P_C \quad CO
\end{array}
$$

III

The complex is moderately soluble in benzene, chloroform, and dichloromethane.

H. DICARBONYLDIHYDRIDOBIS(TRIPHENYLPHOSPHINE)-OSMIUM(II)

$$Na_2OsCl_6 \cdot 6H_2O + 2(C_6H_5)_3P \xrightarrow{\text{KOH, HCHO}} OsH_2(CO)_2[P(C_6H_5)_3]_2$$

Dicarbonyldihydridobis(triphenylphosphine)osmium(II) was first prepared by allowing $OsH_2(CO)_4$ to react with three equivalents of triphenylphosphine in tetrahydrofuran at reflux temperature for 12 hours.[9] It has also been prepared by treatment of $Os(CO)_3$-$[P(C_6H_5)_3]_2$ with hydrogen at 120 atmospheres pressure and 130° in tetrahydrofuran for 12 hours.[9] The yield was 50%, and the product was purified by fractional crystallization from n-heptane-tetrahydro-furan. The procedure described below[2] is a rapid, single-stage synthesis.

Procedure

Solutions of 1.18 g. (2.0 mmoles) of sodium hexachloroosmate(IV) 6-hydrate in 30 ml. of 2-methoxyethanol, aqueous formaldehyde (55 ml., 40% w/v solution), and 2.0 g. of potassium hydroxide in 20 ml. of 2-methoxyethanol are added rapidly and successively to a vigorously stirred, boiling solution of 0.92 g. (3.5 mmoles) of triphenylphosphine in 100 ml. of 2-methoxyethanol. The very dark reaction mixture is heated under reflux for 30 minutes and slowly lightens in color. It is then allowed to cool to room temperature. The precipitated product is filtered, washed successively with ethanol, water, ethanol, and *n*-hexane, and dried *in vacuo*. Yield is 0.93 g. (57%). *Anal.* Calcd. for $C_{38}H_{32}O_2OsP_2$: C, 59.06; H, 4.17; P, 8.02. Found: C, 59.16; H, 4.48; P, 8.25.

Properties

Dicarbonyldihydridobis(triphenylphosphine)osmium(II) forms creamy-white microcrystals which melt at 243–244° in air and at 257–258° in a capillary sealed under nitrogen. The infrared spectrum shows bands at 1926 (m) and 1871 (s) cm.$^{-1}$, attributed to $v(OsH)$, and 2020 (s) and 1990 (vs) cm.$^{-1}$, attributed to $v(CO)$. The high-field n.m.r. pattern (benzene solution) comprises a triplet indicative of stereochemistry IV.

$$\text{IV} \quad \underset{\underset{P_B}{|}\diagup OC}{OC} \!-\!\!\overset{\overset{P_B \diagup H_A}{|}}{Os} \!-\! H_A \qquad H_A \quad \tau 17.08 \qquad J(H_A P_B) \quad 22.75 \text{ Hz.}$$

The complex is moderately soluble in benzene, chloroform, dichloromethane, and tetrahydrofuran.

I. TETRAHYDRIDOTRIS(TRIPHENYLPHOSPHINE) OSMIUM(IV)

$$Na_2OsCl_6 \cdot 6H_2O + 3(C_6H_5)_3P \xrightarrow{NaBH_4} OsH_4[P(C_6H_5)_3]_3$$

Tetrahydridotris(triphenylphosphine)osmium(IV) has been prepared by reaction of sodium hexachloroosmate(IV) or ammonium hexabromoosmate(II) with triphenylphosphine and sodium tetra-hydroborate in boiling ethanol.[1,20]

Procedure

Solutions of 0.57 g. (1.0 mmole) of sodium hexachloroosmate(IV) 6-hydrate in 20 ml. of ethanol (filtered cold extract), and 0.2 g. of sodium tetrahydroborate in 20 ml. of ethanol are added rapidly and successively to a vigorously stirred, boiling solution of 1.57 g. (6.0 mmoles) of triphenylphosphine in 80 ml. of ethanol. Vigorous frothing occurs. The pink reaction solution is heated under reflux for 15 minutes, during which period a white precipitate appears. The precipitated product is filtered, washed successively with ethanol, water, ethanol, and *n*-hexane, and dried *in vacuo*. Yield is 0.95 g. (95%). *Anal.* Calcd. for $C_{54}H_{49}OsP_3$: C, 66.11; H, 5.03; P, 9.47. Found: C, 65.86; H, 5.21; P, 9.18.

Properties

Tetrahydridotris(triphenylphosphine)osmium(IV) forms creamy-white microcrystals which melt at 172–175° in air and at 219–221° in a capillary sealed under nitrogen. The infrared spectrum shows bands at 2086 (w), 2025 (m), 1951 (m), and 1891 (s) cm.$^{-1}$ attributed to v(OsH). The high-field n.m.r. spectrum (benzene solution) comprises a 1:3:3:1 quartet [τ17.85, J(HP) 9 Hz.] which has been interpreted in terms of a nonrigid structure. The complex is sparingly soluble in benzene and dichloromethane.

J. TRICHLORONITROSYLBIS(TRIPHENYLPHOSPHINE)-OSMIUM(II)

$$Na_2OsCl_6 \cdot 6H_2O + 2(C_6H_5)_3P + p\text{-}TolSO_2N(NO)(CH_3) \longrightarrow$$

$$OsCl_3(NO)[P(C_6H_5)_3]_2$$

Trichloronitrosylbis(triphenylphosphine)osmium(II) has been made by a multistage synthesis[21] and also by reaction of sodium hexachloroosmate, nitric oxide, and triphenylphosphine.[22] The procedure[1] eliminates use of gaseous nitric oxide.

Procedure

Solutions of 1.12 g. (2.0 mmoles) of sodium hexachloroosmate(IV) 6-hydrate in 40 ml. of 2-methoxyethanol and 0.84 g. of *N*-methyl-*N*-nitroso-*p*-toluenesulfonamide in 40 ml. of 2-methoxyethanol are added rapidly and successively to a vigorously stirred, boiling solution of 3.16 g. (12 mmoles) of triphenylphosphine in 80 ml. of 2-methoxyethanol. The initially green solution is heated under reflux for 10 minutes and cooled to room temperature. The brown product precipitates and is filtered, washed with ethanol, water, ethanol, and *n*-hexane, and dried *in vacuo*. Yield is 1.12 g. (66%). *Anal.* Calcd. for $C_{36}H_{30}Cl_3NOOsP_2$: C, 50.80; H, 3.55; N, 1.65; P, 7.28; Cl, 12.5. Found: C, 51.11; H, 3.83; N, 1.49; P, 7.12; Cl, 13.1.

Properties

Trichloronitrosylbis(triphenylphosphine)osmium(II) forms brown microcrystals which melt at 265–290° in air and at 330° in a capillary sealed under nitrogen. The infrared spectrum shows a distinctive band at 1850 (vs) cm.$^{-1}$ attributed to $v(NO)$. The complex is slightly soluble in benzene and dichloromethane.

K. HYDRIDOTETRAKIS(TRIPHENYLPHOSPHINE)-RHODIUM(I)

$$RhCl_3 \cdot 3H_2O + 3(C_6H_5)_3P \xrightarrow{\text{KOH}} RhH[P(C_6H_5)_3]_4$$

Hydridotetrakis(triphenylphosphine)rhodium(I) has previously been prepared by addition of triphenylphosphine to preformed hydridotris(triphenylphosphine)rhodium(I) in toluene solution[23,24] and by reaction of preformed chlorotris(triphenylphosphine)-rhodium(I) with hydrazine and hydrogen in an ethanol-benzene medium containing excess triphenylphosphine.[23] Other syntheses employ aluminum alkyls,[25] Grignard reagents,[26] sodium prop-oxide,[27] and hydrogen under pressure[28] as reductants.

The procedure[1,2] given below affords an efficient, one-step synthesis of hydridotetrakis(triphenylphosphine)rhodium(I) from hydrated rhodium trichloride.

Procedure

Hot solutions of 0.26 g. (1.0 mmole) of rhodium trichloride 3-hydrate in 20 ml. of ethanol and 0.4 g. of potassium hydroxide in 20 ml. of ethanol are added rapidly and successively to a vigorously stirred, boiling solution of 2.62 g. (10 mmoles) of triphenylphosphine in 80 ml. of ethanol. The mixture is heated under reflux for 10 minutes and allowed to cool to 30°. The precipitated product is filtered, washed with ethanol, water, ethanol, and *n*-hexane, and dried *in vacuo*. Yield 1.10 g. (97% based on $RhCl_3 \cdot 3H_2O$). *Anal.* Calcd. for $C_{72}H_{61}$-P_4Rh: C, 75.0; H, 5.33; P, 10.73. Found: C, 74.69; H, 5.31; P, 10.52.

Properties

Hydridotetrakis(triphenylphosphine)rhodium(I) forms yellow microcrystals which melt at 145–147° in air and at 154–156° in a capillary sealed under nitrogen. The infrared spectrum shows a band at 2156 (m) cm.$^{-1}$ attributable to $v(RhH)$. The complex is soluble in benzene, chloroform, and dichloromethane.

L. CARBONYLHYDRIDOTRIS(TRIPHENYLPHOSPHINE)-RHODIUM(I)

$$RhCl_3 \cdot 3H_2O + 3(C_6H_5)_3P \xrightarrow{\text{KOH, HCHO}} RhH(CO)[P(C_6H_5)_3]_3$$

Carbonylhydridotris(triphenylphosphine)rhodium(I) was first prepared from $RhCl(CO)[P(C_6H_5)_3]_2$ by reduction with hydrazine in ethanolic suspension.[29] More recent syntheses involve reaction of $RhCl(CO)[P(C_6H_5)_3]_2$ with sodium tetrahydroborate[30] or triethylamine and hydrogen in ethanol[30] containing excess triphenylphosphine. Addition of ethanolic rhodium trichloride solution, aqueous formaldehyde, and ethanolic sodium tetrahydroborate to a boiling solution of triphenylphosphine in ethanol has also been employed to synthesize $RhH(CO)[P(C_6H_5)_3]_3$.[31] The following single-stage procedure[2] utilizes ethanolic potassium hydroxide in place of sodium tetrahydroborate.

Procedure

A solution of 0.26 g. (1.0 mmole) rhodium trichloride 3-hydrate in 20 ml. of ethanol is added to a vigorously stirred, boiling solution of 2.64 g. (10 mmoles) of triphenylphosphine in 100 ml. of ethanol. After a delay of 15 seconds, aqueous formaldehyde (10 ml., 40% w/v solution) and a solution of 0.8 g. of potassium hydroxide in 20 ml. of hot ethanol are added rapidly and successively to the vigorously stirred, boiling reaction mixture. The mixture is heated under reflux for 10 minutes and then allowed to cool to room temperature. The bright yellow, crystalline product is filtered, washed with ethanol, water, ethanol, and *n*-hexane, and dried *in vacuo*. Yield is 0.85 g. (94% based on $RhCl_3 \cdot 3H_2O$). *Anal.* Calcd. for $C_{55}H_{46}OP_3Rh$: C, 71.90; H, 5.05; P, 10.11. Found: C, 72.11; H, 5.17; P, 9.86.

Properties

Carbonylhydridotris(triphenylphosphine)rhodium(I) forms yellow microcrystals which melt at 120–122° in air and at 172–174° in a capillary tube under nitrogen. The infrared spectrum shows bands at 2041 (s) cm.$^{-1}$, attributed to $v(RhH)$, and at 1918 (vs), attributed to $v(CO)$. The high-field n.m.r. spectrum comprises a single signal at $\tau 19.30$, broadened by ligand dissociation and exchange processes. The complex is soluble in benzene, chloroform, and dichloromethane.

M. DICHLORONITROSYLBIS(TRIPHENYLPHOSPHINE)-RHODIUM(I)

$$RhCl_3 \cdot 3H_2O + 2(C_6H_5)_3P + p\text{-TolSO}_2N(NO)(CH_3) \longrightarrow$$

$$RhCl_2(NO)[P(C_6H_5)_3]_2$$

Dichloronitrosylbis(triphenylphosphine)rhodium(I) was first prepared from $[RhCl(NO)_2]_x$ and triphenylphosphine.[32] It has subsequently been isolated by reacting chlorotris(triphenylphosphine)-rhodium(I) with nitric oxide in chloroform solution[33] and by adding triphenylphosphine to an ethanolic solution of "$RhCl_3$-(NO)."[34]

The following efficient single-stage procedure[1] avoids the use of nitric oxide.

Procedure

Solutions of 0.39 g. (1.5 mmoles) of rhodium trichloride 3-hydrate in 30 ml. of ethanol and of 0.63 g. of N-methyl-N-nitroso-p-toluene-sulfonamide in 30 ml. of ethanol are added rapidly and successively to a vigorously stirred, boiling solution of 3.93 g. (15 mmoles) of triphenylphosphine in 120 ml. of ethanol. The mixture is heated under reflux for 10 minutes and then cooled to room temperature. The precipitated brown product is filtered, washed with ethanol, water, ethanol, and n-hexane, and dried *in vacuo*. Yield is 1.08 g. (100% based on $RhCl_3 \cdot 3H_2O$). *Anal.* Calcd. for $C_{36}H_{30}Cl_2NOP_2$-Rh: C, 59.36; H, 4.15; N, 1.92; Cl, 9.73; P, 8.50. Found: C, 59.70; H, 4.19; N, 1.79; Cl, 9.92; P, 8.82.

Properties

Dichloronitrosylbis(triphenylphosphine)rhodium(I) forms light-brown microcrystals which melt at 223–226° in air and at 278–279° in a capillary tube sealed under nitrogen. The infrared spectrum shows a band at 1630 (vs) cm.$^{-1}$ attributed to $v(NO)$. The complex is slightly soluble in benzene, chloroform, and dichloromethane.

N. NITROSYLTRIS(TRIPHENYLPHOSPHINE)RHODIUM

$$RhCl_3 \cdot 3H_2O + 3(C_6H_5)_3P + p\text{-}TolSO_2N(NO)(CH_3) \xrightarrow{KOH}$$

$$Rh(NO)[P(C_6H_5)_3]_3$$

Nitrosyltris(triphenylphosphine)rhodium has previously been prepared by sodium-amalgam reduction of a mixture of rhodium nitrosyl chloride and triphenylphosphine in tetrahydrofuran.[32] Preparations from hydridotetrakis(triphenylphosphine)rhodium(I) and nitric oxide,[24] and from rhodium trichloride, nitric oxide, triphenylphosphine, and zinc dust have also been reported.[35] The following synthesis[1] is convenient and rapid.

Procedure

Solutions of 0.26 g. (1.0 mmole) of rhodium trichloride 3-hydrate in 20 ml. of ethanol of 0.4 g. of N-methyl-N-nitroso-p-toluenesulfonamide in 20 ml. of ethanol and 0.4 g. of potassium hydroxide in 20 ml. of ethanol are added in rapid succession to a well-stirred solution of 2.62 g. (10.0 mmoles) of triphenylphosphine in 80 ml. of boiling ethanol. This mixture is heated under reflux for 10 minutes, cooled to 30°, and then filtered. The red precipitate is washed successively with ethanol., water, ethanol, and *n*-hexane, and dried *in vacuo*. Yield is 0.8 g. (87% based on $RhCl_3 \cdot 3H_2O$). *Anal.* Calcd. for $C_{54}H_{45}NOP_3Rh$: C, 70.52; H, 4.93; N, 1.52; P, 10.1. Found: C, 70.44; H, 4.76; N, 1.4; P, 9.84.

Properties

Nitrosyltris(triphenylphosphine)rhodium forms bright crimson-red microcrystals. It softens at 96°, then melts with decomposition at 160° in air; sealed under nitrogen it melts at 205–206°. The infra-red spectrum shows a peak at 1610 cm.$^{-1}$ (vs) attributed to $v(NO)$. The complex is soluble in dichloromethane, chloroform, and benzene.

O. DICHLORONITROSYLBIS(TRIPHENYLPHOSPHINE)-IRIDIUM(I)

$$Na_2IrCl_6 \cdot 6H_2O + 3(C_6H_5)_3 + p\text{-}TolSO_2(NO)(CH_3) \longrightarrow$$

$$IrCl_2(NO)[P(C_6H_5)_3]_2$$

Dichloronitrosylbis(triphenylphosphine)iridium(I) was first prepared by treating $[Ir(NO)_2\{P(C_6H_5)_3\}_2]ClO_4$ with hydrogen chloride[37] and has subsequently been obtained by a similar reaction involving $Ir(NO)[P(C_6H_5)_3]_3$.[38] Reactions of triphenylphosphine with preformed $IrCl_2(NO)(C_8H_{12})$ in boiling acetone, or with hydrated iridium trichloride and nitric oxide in boiling methanol also afford dichloronitrosylbis(triphenylphosphine)iridium(I).[33] The following procedure[1] is quick and efficient and eliminates the use of nitric oxide.

Procedure

Solutions of 1.40 g. (2.5 mmoles) of sodium hexachloroiridate(IV) 6-hydrate in 50 ml. of 2-methoxyethanol and of 1.05 g. of *N*-methyl-*N*-nitroso-*p*-toluenesulfonamide in 50 ml. of the same solvent are added rapidly and successively to a vigorously stirred, boiling solution of 3.95 g. (15 mmoles) of triphenylphosphine in 100 ml. of 2-methoxyethanol. The mixture is stirred and heated under reflux for 5 minutes, then allowed to cool to room temperature. The precipitated product is filtered, washed with ethanol, water, ethanol, and *n*-hexane, and dried *in vacuo*. Yield is 1.05 g. (52%). *Anal.* Calcd. for $C_{36}H_{30}Cl_2NOP_2Ir$: C, 52.88; H, 3.70; Cl, 8.67; N, 1.71; P, 7.58. Found: C, 53.4; H, 3.73; Cl, 8.46; N, 1.64; P, 7.60.

Properties

Dichloronitrosylbis(triphenylphosphine)iridium(I) forms orange-yellow microcrystals which melt at 247–252° in air and at 308–309° in a capillary tube sealed under nitrogen. The infrared spectrum shows a band at 1560 (vs) cm.$^{-1}$ attributed to $v(NO)$. The complex is moderately soluble in benzene, chloroform, and dichloromethane.

References

1. J. J. Levison and S. D. Robinson, *J. Chem. Soc. (A)*, **1970**, 2947.
2. N. Ahmad, S. D. Robinson, and M. F. Uttley, *ibid.*, *Dalton Trans.*, **1972**, 843.
3. D. Evans, J. A. Osborn, and G. Wilkinson, *Inorganic Syntheses*, **11**, 99 (1968).
4. J. Chatt and B. L. Shaw, *Chem. Ind.*, **1961**, 290.
5. L. Vaska and J. W. DiLuzio, *J. Am. Chem. Soc.*, **83**, 1262 (1961) and references therein.
6. P. S. Hallman, B. R. McGarvey, and G. Wilkinson, *J. Chem. Soc. (A)*, **1968**, 3143.
7. J. P. Collman and W. R. Roper, *J. Am. Chem. Soc.*, **87**, 4008 (1965).
8. F. Piacenti, M. Bianchi, E. Benedetti, and G. Sbrana, *J. Inorg. Nucl. Chem.*, **29**, 1389 (1967).
9. F. L'Eplattenier and F. Calderazzo, *Inorg. Chem.*, **7**, 1290 (1968).
10. D. Rose, J. D. Gilbert, R. P. Richardson, and G. Wilkinson, *J. Chem. Soc. (A)*, **1969**, 2610.
11. M. B. Fairy and R. J. Irving, *ibid. (A)*, **1966**, 475.
12. K. R. Laing and W. R. Roper, *Chem. Commun.*, **1968**, 1556.
13. M. H. B. Stiddard and R. E. Townsend, *ibid.*, **1969**, 1372.
14. J. J. Levison and S. D. Robinson, *Chem. Ind.*, **1969**, 1514.
15. T. I. Eliades. R. O. Harris, and M. C. Zia, *Chem. Commun.*, **1970**, 1709.
16. K. R. Grundy, K. R. Laing, and W. R. Roper, *ibid.*, **1970**, 1500.
17. L. Vaska, *Z. Naturforsch.*, **15B**, 56 (1960).
18. L. Vaska, *J. Am. Chem. Soc.*, **86**, 1943 (1964).
19. L. Vaska, *ibid.*, **88**, 4100 (1966).
20. G. J. Leigh, J. J. Levison, and S. D. Robinson, *Chem. Commun.*, **1969**, 705.

21. K. R. Laing and W. R. Roper, *ibid.*, **1968,** 1968.
22. A. Aràneo, V. Valenti, and F. Cariati, *J. Inorg. Nucl. Chem.*, **32,** 1877 (1970).
23. K. C. Dewhirst, W. Keim, and C. A. Reilly, *Inorg. Chem.*, **7,** 546 (1968).
24. B. Ilmaier and R. S. Nyholm, *Naturwiss.*, **56,** 415 (1969).
25. A. Yamamoto, S. Kitazume, and S. Ikeda, *J. Am. Chem. Soc.*, **90,** 1089 (1968).
26. M. Takesada, H. Yamazaki, and N. Hagihara, *J. Chem. Soc. Japan,* **89,** 1121 (1968).
27. G. Gregorio, G. Pregaglia, and R. Ugo, *Inorg. Chim. Acta*, **3,** 89 (1969).
28. M. Takesada, H. Yamazaki, and N. Hagihara, *J. Chem. Soc. Japan*, **89,** 1126 (1968).
29. S. S. Bath and L. Vaska, *J. Am. Chem. Soc.*, **85,** 3500 (1963).
30. P. S. Hallman, D. Evans, J. A. Osborn, and G. Wilkinson, *Chem. Commun.*, **1967,** 305.
31. D. Evans, G. Yagupsky, and G. Wilkinson, *J. Chem. Soc. (A)*, **1968,** 2660.
32. W. Hieber and K. Heinicke, *Z. Anorg. Chem.*, **316,** 321 (1962).
33. G. R. Crooks and B. F. G. Johnson, *J. Chem. Soc. (A)*, **1970,** 1662.
34. M. C. Baird, *Inorg. Chim. Acta*, **5,** 46 (1971).
35. J. P. Collman, N. W. Hoffman, and D. E. Morris, *J. Am. Chem. Soc.*, **91,** 5659 (1969).
36. R. G. Hayter, *ibid.*, **83,** 1259 (1961).
37. L. Malatesta, M. Angoletta, and G. Caglio, *Angew. Chem. Int. Ed.*, **2,** 739 (1963).
38. C. A. Reed and W. R. Roper, *Chem. Commun.*, **1969,** 155.

14. FLUORO COMPLEXES OF RHODIUM(I) AND IRIDIUM(I)

Submitted by L. VASKA* and J. PEONE, JR.*,†
Checked by E. R. WONCHOBA‡ and G. W. PARSHALL‡

The compounds described here, *trans*-$[MF(CO)(Ph_3P)_2]$, in which M = Ir or Rh,[1-3] are useful as starting materials for a one-step convenient preparation of a variety of compounds of these metals:[2]

$$trans\text{-}[MF(CO)(Ph_3P)_2] + M'A \longrightarrow$$

$$trans[MA(CO)(Ph_3P)_2] + M'F$$

*Department of Chemistry, Clarkson College of Technology, Potsdam, N.Y. 13676.
†Present address: Department of Chemistry, Broome Community College, Binghamton, N.Y. 13902.
‡Central Research Department, Experimental Station, E. I. du Pont de Nemours & Company, Wilmington, Del. 19898.

where M = Ir, Rh; M' = Li, Na, or NH_4; A = Cl, Br, I, NCO, NCS, NCSe, N_3, $N(CN)_2$, CN, NO_2, NO_3, $OClO_3$, OH, OPh, SePh, OOCH, $OOCCH_3$, or OOCPh.

These derivatives provide a unique opportunity to compare the ligational properties of a large number of anions through a study of their electronic and vibrational spectra.[2] The *new* d^8 complexes are of further interest in that they undergo reversible addition reactions with covalent molecules, similar to those of their well-known halogeno analogs.[4] According to their x-ray powder patterns, the *trans*-[$MF(CO)(Ph_3P)_2$] compounds are isomorphous with their precursors, *trans*-[$MCl(CO)(Ph_3P)_2$].[5-7]

The rhodium complex, *trans*-[$RhF(CO)(Ph_3P)_2$], was first synthesized by Grinberg *et al.*[1] by the reaction of *trans*-[$RhCl(CO)(Ph_3P)_2$] with silver fluoride in boiling acetone; a yield of *ca.* 60% was reported. The method described here affords the rhodium complex in *ca.* 90% yields and the analogous iridium compound, *trans*-[$IrF(CO)(Ph_3P)_2$][2], in 80% yield. The Ir complex has also been obtained by the reaction of [$Ir(CO)(CH_3CN)(Ph_3P)_2$]$^+$ with "fluoride."[3]

A. *trans*-CARBONYLFLUOROBIS(TRIPHENYLPHOSPHINE)-RHODIUM(I)

$$2\ trans\text{-}[RhCl(CO)(Ph_3P)_2] + Ag_2CO_3 + 2NH_4F \xrightarrow{\ CH_3OH\ }$$

$$2\ trans\text{-}[RhF(CO)(Ph_3P)_2] + 2AgCl + CO_2 + 2NH_3 + H_2O$$

Procedure

A 1.03-g. (1.49-mmole) sample of powdered *trans*-[$RhCl(CO)$-$(Ph_3P)_2$][7] is added to a 300-ml. saturated solution of ammonium fluoride (10 g., 0.27 mole) in methanol in a 500-ml. Erlenmeyer flask. The resulting yellow suspension/solution is stirred by manual swirling and heated to boiling on a steam bath. After about 15 minutes, all but a small amount of the rhodium complex has dissolved. To this boiling mixture, 0.204 g. (0.745 mmole) of solid silver carbonate (Ag:Rh = 1.0) is added in several small portions over a period of 1 minute; some effervescence (CO_2) is observed during this addition. Boiling of the reaction mixture is continued for three additional minutes, during which time silver carbonate dissolves and silver chloride precipitates. The hot suspension is filtered through a medium-porosity fritted disk (250-ml. filter crucible), under reduced pressure, into 300 ml. of vigorously stirred aqueous solution of 0.1 M ammonium fluoride in a 1-l. filter flask. A white, photo-sensitive residue (AgCl) remains on the filter and a light-yellow precipitate forms as the yellow filtrate mixes with the aqueous NH_4F solution in the filter flask. The yellow product is separated by filtration (30-ml. medium-porosity filter crucible), washed with two 50-ml. portions of water, and dried in a vacuum desiccator for 24 hours. The yield is 0.903 g. (90%). *Anal.* Calcd. for $RhFOP_2$-$C_{37}H_{30}$: Rh, 15.3; F, 2.9; O, 2.4; P, 9.2; C, 64.7; H, 4.5; Cl, 0.0. Found: Rh, 13.3; F, 2.9; O, 3.4; P, 9.0; C, 64.2; H, 4.4; Cl, trace.

Properties

trans-Carbonylfluorobis(triphenylphosphine)rhodium(I) is an air-stable crystalline material. The yellow complex is soluble in benzene, chloroform, acetone, and methanol, is slightly soluble in ethanol and diethyl ether, and is insoluble in water and hexane.

The infrared spectrum shows characteristic absorption bands (cm^{-1}): ν_{CO}, 1956 (s) and 1965 (m) in Nujol (solid-state splitting), 1971 (vs) in $CHCl_3$; ν_{Rh-CO}, 603 (ms); ν_{RhF}, 478 (m) cm.$^{-1}$, all in Nujol; the latter assignment is not certain because of the proximity of the $(C_6H_5)P$ bands. For comparison, the corresponding frequencies in the spectrum of the starting material, *trans*-$[RhCl(CO)(Ph_3P)_2]$, are: ν_{CO}, 1965 (vs) in Nujol, 1980 (vs) in $CHCl_3$; ν_{Rh-CO}, 576 (ms) in Nujol. These infrared spectral differences, together with the absence of the ν_{RhCl} band, 315 (m), in the spectrum of *trans*-$[RhF-(CO)(Ph_3P)_2]$, provide a convenient means for judging the purity of the fluoro complex.

The visible spectrum may also be used for a rapid characterization of *trans*-$[RhF(CO)(Ph_3P)_2]$: in benzene solution, the complex exhibits one absorption band at 358 nm. (λ_{max}), $a = 4000\ M^{-1}$ cm.$^{-1}$. The spectrum of its chloro analog shows $\lambda_{max} = 367$ nm., $a = 3780$ M^{-1} cm.$^{-1}$. The solutions of the molecular fluoro complex $[(2-5) \times 10^{-4}M, 25°]$ in acetone and nitromethane are nonconducting:

$\Lambda_M = 0.0$ and 4.4, respectively; in methanol, however, $\Lambda_M = 35 \ \Omega^{-1}$, which points to a partial solvolytic ionization.[2]

B. *trans*-CARBONYLFLUOROBIS(TRIPHENYLPHOSPHINE)-IRIDIUM(I)

$$2 \text{ } trans\text{-}[IrCl(CO)(Ph_3P)_2] + Ag_2CO_3 + 2NH_4F \xrightarrow{CH_3OH}$$

$$2 \text{ } trans\text{-}[IrF(CO)(Ph_3P)_2] + 2AgCl + 2NH_3 + H_2O$$

Procedure

This complex is synthesized by a procedure similar to the one described above for the rhodium compound, *trans*-[RhF(CO)-(Ph₃P)₂], *except* that when the addition of silver carbonate is completed (1 minute), boiling of the reaction mixture prior to filtration is continued for as long as is necessary to dissolve the Ag_2CO_3, 1 to 5 minutes, *but no longer*. Longer heating times result in metallic deposits and lower yields of the product.

Quantities used: 1.07 g. (1.37 mmoles) of *trans*-[IrCl(CO)(Ph₃P)₂][8], 0.188 g. (0.685 mmole) of Ag_2CO_3. A yield of 0.803 g. (80%) is obtained. *Anal.* Calcd. for $IrFOP_2C_{37}H_{30}$: F, 2.5; P, 8.1; C, 58.0; H, 3.9. Found: F, 2.6; P, 8.0; C, 57.7; H, 4.2.

Properties

trans-Carbonylfluorobis(triphenylphosphine)iridium(I) is a crystalline yellow solid. It is somewhat air-sensitive and must be stored in an inert atmosphere (N_2, Ar) or *in vacuo*. The complex shows the same solubility characteristics as the corresponding rhodium compound. The purity of *trans*-[IrF(CO)(Ph₃P)₂] and its distinctive properties are reflected by its vibrational and electronic spectra: ν_{CO}, 1942 (vs) in Nujol and 1957 (vs) in CHCl₃; ν_{Ir-CO}, 629 (ms) and ν_{IrF}, 451 (ms) cm.$^{-1}$, both in Nujol; λ_{max} in benzene: 427, 378, and 335 nm., $a = 815, 4300$, and 3740 M^{-1} cm.$^{-1}$, respectively. For the chloro complex, the corresponding values are: ν_{CO}, 1956 (vs) in Nujol and 1965 (vs) in CHCl₃; ν_{Ir-CO}, 603 (ms) and ν_{IrCl}, 317 (m) cm.$^{-1}$, both in Nujol; λ_{max} in benzene $= 439, 387$, and 339 nm.; $a = 731, 4040$, and 3160 M^{-1} cm.$^{-1}$, respectively. Molar conductances for $(2-5) \times 10^{-4}$ M solutions of *trans*-$[IrF(CO)(Ph_3P)_2]$ at $25°$: Λ_M (Ω^{-1}) = 0.2 in acetone, 9.3 in nitromethane, and 22 in methanol.[2]

References

1. A. A. Grinberg, M. M. Singh, and Yu. S. Varshavskii, *Russ. J. Inorg. Chem.*, **13**, 1399 (1968).
2. L. Vaska and J .Peone, Jr., *Chem. Commun.*, **1971**, 418.
3. G. R. Clark, C. A. Reed, W. R. Roper, B. W. Skelton, and T. N. Waters, *ibid.*, **1971**, 758.
4. L. Vaska, *Accounts Chem. Res.*, **1**, 335 (1968); L. Vaska, *Inorg. Chim. Acta*, **5**, 295 (1971); L. Vaska and M. F. Werneke, *Trans. N.Y. Acad. Sci.*, **31**, 70 (1971); L. Vaska, L. S. Chen, and C. V. Senoff, *Science*, **174**, 587 (1971).
5. L. Vallarino, *J. Chem. Soc.*, **1957**, 2287.
6. L. Vaska and J. W. DiLuzio, *J. Am. Chem. Soc.*, **83**, 2784 (1961).
7. D. Evans, J. A. Osborn, and G. Wilkinson, *Inorganic Syntheses*, **11**, 99 (1968).
8. K. Vrieze and J. P. Collman, C. T. Sears, and M. Kuboto, *ibid.*, **11**, 101 (1968).

15. COVALENT PERCHLORATO COMPLEXES OF IRIDIUM AND RHODIUM

Submitted by J. PEONE, JR.,*·† BRENDAN R. FLYNN,* and L. VASKA*
Checked by W. H. BADDLEY‡ and D. S. HAMILTON‡

Tetrahedral perchlorate ion, $[ClO_4]^-$, is an uncommon ligand.[1] The molecular perchlorato complexes of iridium and rhodium[2] described here are thus of inherent interest as such, but their principal importance lies in their versatile reactivity. These compounds undergo addition, substitution, and addition-substitution reactions with many molecules and ions.[3] In particular, the latter conversions lead to a remarkable number of cationic d^8 complexes of these metals, which offer themselves as a unique series for a study of the electronic properties of a variety of molecules as ligands (L).[3] Not less significant are the substitution reactions in which the perchlorate ligand is replaced by other unusual anions.[4]

*Department of Chemistry, Clarkson College of Technology, Potsdam, N.Y. 13676.
†Present address: Department of Chemistry, Broome Community College, Binghamton, N.Y. 13902.
‡Department of Chemistry, Louisiana State University, Baton Rouge, La. 70803.

The present preparative procedure for perchlorato complexes of rhodium and iridium appears to be of general utility for the synthesis of perchlorato complexes of transition metals. The assignment of trans planar structures for $[M(OClO_3)(CO)(Ph_3P)_2]$ (M = Ir, Rh) is based on the vibrational and electronic spectral similarities between the perchlorato complexes and their well-known chloro precursors, *trans*-$[MCl(CO)(Ph_3P)_2]$.[5,6]

■ **Caution.** *Although no explosions have occurred with these perchlorato complexes, these compounds are potentially very hazardous. Adequate shielding and protective clothing should be worn when handling them.*[7]

A. *trans*-CARBONYLPERCHLORATOBIS(TRIPHENYL-PHOSPHINE)IRIDIUM(I)

$$trans\text{-}[IrCl(CO)(Ph_3P)_2] + AgClO_4 \xrightarrow{C_6H_6}$$

$$trans\text{-}[Ir(OClO_3)(CO)(Ph_3P)_2] + AgCl$$

Procedure

A solution of 2.41 g. (3.09 mmoles) of *trans*-$[IrCl(CO)(Ph_3P)_2]$[6,8] in 200 ml. of benzene* is prepared in a 250-ml. Erlenmeyer flask containing a magnetic stirring bar. The flask is wrapped in aluminum foil to exclude light. The solution is stirred and deaerated by a stream of nitrogen passed through the solution for 10 minutes. Under continuing flow of nitrogen, solid silver perchlorate, 0.637 g. (3.08 mmoles, Ag:Ir = 1.0.■ **Caution.** *Excess AgClO₄ should be avoided because its solutions in benzene are dangerously explosive!*) is added to the solution. The flask is quickly stoppered, and the mixture is stirred until all the AgClO₄ dissolves. The dissolution requires 15–60 minutes, depending on the size of the silver perchlorate crystals used. The disappearance of AgClO₄ can be readily determined visually because its particles are much larger and more irregular than the fine microcrystalline silver chloride which is produced. The resulting deep yellow-orange solution containing

---

*The starting complex is not completely soluble in this quantity of the solvent. The remaining solid portion dissolves during the reaction to give the more soluble perchlorato complex. Larger volumes of benzene reduce the yield of the solid product.

suspended AgCl is filtered in air through a medium-porosity fritted disk (250-ml. filter crucible) under reduced pressure into 500 ml. of vigorously stirred* hexane in a 1-l. filter flask. Only white AgCl remains on the filter, and a deep-yellow precipitate forms as the filtrate mixes with the hexane.* The product is collected by filtration (45-ml. medium-porosity crucible), washed with two 50-ml. portions of hexane and dried *in vacuo* for 12 hours. The yield is 2.14 g. (82%; yield range: 75–92%). *Anal.* Calcd. for $C_{37}H_{30}ClIrO_5P_2$: C, 52.6; H, 3.6; Cl, 4.2; Ir, 22.8; O, 9.5; P, 7.35. Found: C, 52.5; H, 4.2; Cl, 5.0; Ir, 21.9; O, 9.2; P, 6.8.

The final benzene-hexane filtrate still contains some dissolved *trans*-[Ir(OClO$_3$)(CO)Ph$_3$P)$_2$]. On addition of triphenylphosphine (10 g., 39 mmoles) to this solution, an orange precipitate of the *ionic perchlorate*, [Ir(CO)(Ph$_3$P)$_3$][ClO$_4$],[2] is obtained which is collected and treated as above; yield, 0.70 g. (17%). This procedure thus affords a nearly quantitative (99%) recovery of the total iridium used.

Properties

The yellow *trans*-carbonylperchloratobis(triphenylphosphine)iridium(I) is a crystalline solid which slowly decomposes on standing in air. It must be stored in an inert atmosphere (N$_2$, Ar) or *in vacuo*, but even under these conditions the complex deteriorates within several weeks. For studies of its properties and reactions, freshly prepared samples should be used. The compound is insoluble in water and hexane, and soluble in chloroform, benzene, ethanol, acetone, methanol, nitromethane, and chlorobenzene. In the latter four solvents, the molar conductances Λ_M of (2–5) $\times$ 10^{-4} M solutions at 25° are: 160, 110, 77, and 0.7 Ω^{-1}, respectively, which shows that a solvolytic ionization occurs in polar solvents.[2]

The complex is conveniently identified and its purity judged by its characteristic infrared spectrum ν_{CO}, 1985 (vs) and 1978 (vs) in Nujol (solid-state splitting), 1990(vs) in CHCl$_3$; ν_{Ir-CO}, 585(m); ν_{OClO_3}, 1160(m), 1130(s), 1050(ms), 920(w), and 620(ms) cm.$^{-1}$, all in Nujol. The visible spectrum in benzene solution at 25°C,

---

*Stirring of this mixture is important; otherwise the crystals of the forming product tend to adhere to the wall of the filter flask and are difficult to remove for the second filtration.

λ_{max} = 440, 387, and 327 nm.; a = 590, 2650, and 3080 M^{-1}cm.$^{-1}$, respectively. For comparison with the spectral data of the starting material, *trans*-[IrCl(CO)(Ph$_3$P)$_2$], see "Properties of *trans*-[IrF-(CO)(Ph$_3$P)$_2$]."[10]

B. *trans*-CARBONYLPERCHLORATOBIS-(TRIPHENYLPHOSPHINE)RHODIUM(I)

trans-[RhCl(CO)(Ph$_3$P)$_2$] + AgClO$_4$ $\xrightarrow{\text{C}_6\text{H}_6}$

$$\text{\textit{trans}-[Rh(OClO}_3\text{)(CO)(Ph}_3\text{P)}_2\text{]} + \text{AgCl}$$

Procedure

This compound is prepared by exactly the same procedure as described above for the iridium complex, *trans*-[Ir(OClO$_3$)(CO)-(Ph$_3$P)$_2$] (Procedure A). Quantities used: 2.22 g. (3.22 mmoles) of *trans*-[RhCl(CO)(Ph$_3$P)$_2$];[9] 0.672 g. (3.24 mmoles, Ag:Rh = 1.0) of AgClO$_4$. The yield is 2.32 g. (95%; yield range 92–98%). *Anal.* Calcd. for C$_{37}$H$_{30}$ClO$_5$P$_2$Rh: C, 58.5; H, 4.0; Cl, 4.7; P, 8.2. Found: C, 58.1; H, 4.3; Cl, 4.9; P, 7.8. Molecular weight: calcd., 754; found in benzene, 776.

Properties

trans-Carbonylperchloratobis(triphenylphosphine)rhodium(I) is a yellow crystalline solid which is stable in air for several months. Its solubility properties are qualitatively the same as cited above for the analogous iridium complex. Molar conductances for (2–5) × 10^{-4} M solutions of *trans*-[Rh(OClO$_3$)(CO)(Ph$_3$P)$_2$] at 25° are: Λ_M = 150, 100, and 75Ω^{-1} in acetone, methanol, and nitromethane, respectively.

The complex is best characterized by its infrared spectrum, which also serves as a means to estimate its purity. ν_{CO}, 1995(vs) and 1985(vs) in Nujol (solid-state splitting), 2000(vs) in CHCl$_3$; ν_{Rh-CO}, 572(m); ν_{OClO_3}, 1160(m), 1130(s), 1070(ms), 920(mw), and 620(m) cm.$^{-1}$, all in Nujol. The visible spectrum in benzene solution at 25°, λ_{max} = 350 nm., a = 3160 M^{-1}cm.$^{-1}$. For comparison with the infrared and visible spectral data for the chloro complex, *trans*-[RhCl(CO)(Ph$_3$P)$_2$], see "Properties of *trans*-[RhF(CO)(Ph$_3$P)$_2$]."[10]

References

1. A. R. Davis, C. J. Murphy, and R. A. Plane, *Inorg. Chem.*, **9**, 423 (1970).
2. J. Peone, Jr., and L. Vaska, *Angew. Chem. Int. Ed.*, **10**, 511 (1971).
3. L. Vaska and J. Peone, Jr., *Suomen Kemistilehti B*, **44**, 317 (1971).
4. L. Vaska and W. V. Miller, *Inorganic Syntheses*, **15**, 72 (1974); L. Vaska and B. R. Flynn, unpublished results.
5. L. Vallarino, *J. Chem. Soc.*, **1957**, 2287.
6. L. Vaska and J. W. DiLuzio, *J. Am. Chem. Soc.*, **83**, 2784 (1961).
7. M. Burton, *Chem. Eng. News*, **28**, 55 (1970).
8. K. Vrieze, J. P. Collman, C. T. Sears, Jr., and M. Kubota, *Inorganic Syntheses*, **11**, 101 (1968).
9. J. A. McCleverty and G. Wilkinson, *Inorganic Syntheses*, **8**, 214 (1966); J. A. Osborn and G. Wilkinson, *ibid.*, **10**, 67 (1967).
10. L. Vaska and J. Peone, Jr., *Inorganic Syntheses*, **15**, 64 (1974).

16. *trans*-CARBONYL(CYANOTRIHYDRIDOBORATO)BIS-(TRIPHENYLPHOSPHINE)RHODIUM(I)

$$trans\text{-}[Rh(OClO_3)(CO)(Ph_3P)_2] + Na[BH_3(CN)] \xrightarrow{C_2H_5OH}$$

$$trans\text{-}[Rh(NCBH_3)(CO)(Ph_3P)_2] + NaClO_4$$

Submitted by L. VASKA* and W. V. MILLER*
Checked by W. H. BADDLEY† and D. S. HAMILTON†

This preparation of a cyanotrihydridoborato (BH_3CN^-) complex of rhodium[1] is of interest in that (1) the BH_3CN^- complexes of Rh and Ir appear to be intermediates in the synthesis of hydrido complexes of these metals,[1] and (2) it illustrates the usefulness of a covalent perchlorato complex[2] as a starting material for synthesis. A similar procedure is used to synthesize other cyanotrihydridoborato and tetrahydridoborato complexes, including *trans*-[Ir(NCBH_3)-(CO)((C_6H_{11})_3P)_2] and *trans*-[Rh(H_2BH_2)(CO)((C_6H_{11})_3P)_2].[1]

The assignment of trans structures for the hydridoborato complexes is based on the vibrational spectral similarities between them and analogous trans-planar derivatives of rhodium and iridium.

*Department of Chemistry, Clarkson College of Technology, Potsdam, N.Y. 13676.
†Department of Chemistry, Louisiana State University, Baton Rouge, La. 70803.

Procedure

A saturated solution of $Na[BH_3(CN)]$* in 5 ml. of absolute ethanol is added to a stirred solution of 0.100 g. (0.133 mmole) of *trans*-$[Rh(OClO_3)(CO)(Ph_3P)_2]^2$ in 10 ml. of absolute ethanol in an open beaker (25°). A yellow precipitate forms immediately which is then separated by filtration through a medium-porosity sintered-glass filter, washed with water (10 ml.), absolute ethanol (10 ml.), and hexane (10 ml.), and dried *in vacuo*. The yield is 0.078 g. (81%). *Anal.* Calcd. for $RhBNOP_2C_{38}H_{33}$: B, 1.55; C, 65.6; H, 4.8. Found: B, 1.99; C, 64.5; H, 4.7.

Properties

The compound is an air-stable yellow solid. It is soluble in benzene and chloroform and insoluble in water, alcohols, and hexane. The purity of the complex can be ascertained by its characteristic infrared spectrum: ν_{BH}, 2360(s, b); δ_{BH}, 1109(m); ν_{CN}, 2192(s); ν_{CO}, 1997(vs); ν_{Rh-CO}, 581(m), all in Nujol; ν_{CO}, 2000 cm.$^{-1}$ in $CHCl_3$. The visible spectrum in benzene solution: $\lambda_{max} = 374$ nm., $a = 4220\ M^{-1}$cm.$^{-1}$.

References

1. L. Vaska, W. V. Miller, and B. R. Flynn, *Chem. Commun.*, **1971**, 1615.
2. J. Peone, Jr., B. R. Flynn, and L. Vaska, *Inorganic Syntheses*, **15**, 72 (1974).

*This material, obtained from Alfa Inorganics, Inc. (P.O. Box 159, Beverly, Mass. 01915), is a yellow-brown solid which gives an odor of HCN; it is used without purification.

OTHER TRANSITION-METAL COMPOUNDS

17. π-ALLYL COMPLEXES OF PALLADIUM AND PLATINUM

Submitted by J. LUKAS*
Checked by B. E. MORRIS† and R. A. CLEMENT†

π-Allyl complexes of palladium are probably more thoroughly investigated than those of any other metal. This preference is partially due to their stability and ease of preparation. Their syntheses were first reported in 1959 by Smidt[1] and Huttel.[2]

The types of reaction leading to π-allylchloropalladium complexes can be divided into two groups:

1. Nucleophilic substitution at an allylic carbon by palladium:

When X = H, π-allylchloropalladium complexes are formed via π-olefin complexes by elimination of hydrogen chloride.[3] The reaction is nonselective insofar as each hydrogen α to the double bond in a monoolefin can be eliminated. The course of the reaction

*Koninklijke/Shell-Laboratorium, Amsterdam, Badhuisweg 3, Netherlands.
†Central Research Department, Experimental Station, E. I. du Pont de Nemours & Company, Wilmington, Del. 19898.

is related to the relative stabilities of the π-allyl complexes produced and is mainly governed by the steric effects of the substituents on the allyl group.

The reaction is an equilibrium that can be shifted to the right by introduction of a base: sodium acetate is one of the reagents of choice with branched olefins.[4] For unbranched alkenes, dimethylformamide is preferred.[5] Inorganic bases have also been used[6] but are less satisfactory. Addition of copper(I) chloride[7] improves the yields, since any Pd(II) that is reduced by the olefin (which in turn is converted into an aldehyde or ketone) is reoxidized.

When X = Cl or OH, the driving force in the reaction is the reduction of an olefin-PdCl$_2$ complex by a reducing agent, the best agents for this purpose being tin(II) chloride[8] and carbon monoxide.[9]

It is safe to assume that this reduction can occur only in the ligand sphere of the metal. Indeed, it has been suggested[9] that the reduction with CO/H_2O proceeds via an intermediate complex containing both carbon monoxide and olefin as ligands. Tin(II) chloride presumably participates as an $[SnCl_3]^-$ ligand.

2. Addition of palladium chloride to conjugated or cumulated dienes:

I   II

The formation of π-allyl complexes from 1, 3-dienes is very general.[9] Its scope has hardly been explored yet. Its main potential lies in the ease of exchange of the substituent in the allylic ligand,[10,11] and in the reversibility of the reaction leading to the complex.[12]

With 1, 2-dienes,[13,14] depending on the relative concentrations of the reagents, a 2-X-π-allyl complex (I) is formed or another molecule of the allene is inserted into the C_2-X bond (II).

The only preparation of a π-allylplatinum complex—aside from the generally applicable reaction of a metal halide with an allyl Grignard reagent—is the reduction of an allyl chloride–$PtCl_2$ mixture with tin(II) chloride.[15]

A. DI-μ-CHLORO-BIS[(1-ETHYL-2-METHYL-π-ALLYL)-PALLADIUM]

$$2NaCl + PdCl_2 \longrightarrow Na_2PdCl_4$$

$$2(CH_3)_2C{=}CHCH_2CH_3 + 2Na_2PdCl_4 + 2NaOOCCH_3 \longrightarrow$$

$$+ 6NaCl + 2CH_3COOH$$

Procedure

Sodium acetate is dried by melting under vacuum or in an anhydrous atmosphere to allow adsorbed water to evaporate.[16] (The checkers used commercial anhydrous sodium acetate.)

Palladium(II) chloride (1.0 g., 5.6 mmoles), 0.66 g. (11.3 mmoles) of sodium chloride, and 0.93 g. (11.3 mmoles) of sodium acetate are dissolved in 100 ml. of glacial acetic acid at 85°. The solution is filtered through a fluted filter paper, and 1.1 ml. (10 mmoles) of 2-methyl-2-pentene is added to the stirred filtrate. The mixture is kept at 85° until the color has completely changed from red to yellow (approx. 30 minutes). The solution is then poured into 500 ml. of H_2O, and the solution is extracted four times with 50-ml. portions of dichloromethane. The combined extracts are consecutively washed with water, aqueous sodium hydrogen carbonate, and water. They are dried over Na_2SO_4. Evaporation of the solvent and drying under vacuum yields 1 g. (80%) of the yellow solid product.

The product can be recrystallized by dissolving it in dichloromethane, adding *n*-pentane until the solution becomes turbid, and cooling to $-80°$ in a Dry Ice–acetone bath. The checkers obtained *ca.* 0.4 g. of recrystallized product. This product contained small amounts of carbonyl-containing impurities as judged by infrared absorption at 1740 and 1920 cm.$^{-1}$. *Anal.* Calcd. for $[C_6H_{11}ClPd]_2$: C, 32.2; H, 4.9; Cl, 15.8; Pd, 47.3. Found: C, 31.8; H, 4.5; Cl, 16.2; Pd, 48.2.

Properties

Di-μ-chloro-bis[(1-ethyl-2-methyl-π-allyl) palladium], like most alkyl- and aryl-substituted π-allylchloropalladium dimers, is a yellow, crystalline, air-stable solid, which is soluble in organic solvents of medium polarity (acetone, dichloromethane, chloroform, benzene). The reactions of these compounds have been described in several reviews.[17]

The ^{1}H n.m.r. spectrum (60 MHz., $CDCl_3$ solution, Me_4Si external standard) shows singlets at $\delta 3.75$ and 2.72 and a triplet ($J = 7$ Hz.) at $\delta 3.62$ assignable to allylic protons and a methyl singlet at $\delta 2.10$ in the appropriate ratio of 1:1:1:3. The ethyl protons produce complex multiplets at higher fields. (In the checker's specimen, these multiplets gave a cumulative intensity of 6 rather than the expected 5.)

B. DI-μ-CHLORO-BIS-{[1-(HYDROXYETHYL)-3-METHYL-π-ALLYL]PALLADIUM}

$2CH_3CH{=}CHCH{=}CHCH_3 + 2PdCl_2(NCC_6H_5)_2 + 2H_2O \longrightarrow$

$$
\left[
\begin{array}{c}
CH_3 \\
| \\
\diagleft\;PdCl \\
| \\
CH{-}OH \\
| \\
CH_3
\end{array}
\right]_2
+ 4C_6H_5CN + 2HCl
$$

Procedure

Bis(benzonitrile)dichloropalladium(II)[18] (3.00 g., 7.8 mmoles) is dissolved in a mixture of 40 ml. of water and 60 ml. of acetone. *trans, trans*-2, 4-Hexadiene (3.5 ml., 35 mmole) is added. After 10 minutes the solution is extracted with dichloromethane, and the extract is washed with water and dried over Na_2SO_4. After evaporation to dryness, the residue is dissolved in a small amount (5–10 ml.) of CH_2Cl_2 and filtered. The filtrate is diluted with pentane until the solution becomes turbid and a precipitate appears. In addition, the solution is cooled to $-80°$. Filtration through a sintered-glass filter and drying under vacuum gives 1.5 g. (80% theoretical) of the title compound.

Any conjugated diene can be used. Water can be replaced by other nucleophilic compounds, e.g., alcohols[14] or acids.[20] If the reaction is run in the absence of a nucleophilic compound, the respective chloro-substituted complex is obtained.

Properties

The compound is stable to air and soluble in $CHCl_3$, CH_2Cl_2, and acetone. The functional group ($-OH$, $-OR$, $-OOCR$) is very labile and solvolyzes easily.[19] The intermediate carbonium ion in the solvolysis may be stabilized as a result of overlap with a filled metal orbital.

The [1]H n.m.r. spectrum (60 MHz., $CDCl_3$ solutions, δ in p.p.m. down field from external TMS standard) showed a triplet at $\delta 5.58$ ($J = 11$ Hz., area = 2, central proton of allyl triad), multiplet absorption around $\delta \approx 4.00$ (area = 6, other protons of the allyl triad and the proton α to the hydroxyl group), a broad doublet at $\delta 2.50$ ($J = 4$ Hz., area = 2, hydroxyl proton), and two doublets at $\delta 1.47$ and 1.35 ($J = 2$ Hz. for both, area = 12, methyl protons).

C. TETRAKIS[ALLYLCHLOROPLATINUM(II)]

$$4[PtCl_4]^{2-} + 4SnCl_2 + 4CH_2{=}CHCH_2Cl \longrightarrow$$

$$[C_3H_5PtCl]_4 + 4[SnCl_6]^{2-}$$

Procedure

Tetrahydrofuran is purified by distillation from $LiAlH_4$.[21] Anhydrous tin(II) chloride* may be obtained by dehydrating tin(II)-chloride 2-hydrate according to a procedure from "Organikum," p. 697, (Deutscher Verlag der Wissenschaften, Berlin, 1969). In a 300-ml. beaker, 102 g. of acetic anhydride and 113 g. of tin(II) chloride 2-hydrate are mixed. Dehydration is so exothermic that the anhydride begins to boil. (■ **Caution.** *Fume hood.*) At the same time, the anhydrous salt precipitates. After 90 minutes it is filtered off and washed twice with 50 ml. of dry ether.

In a 100-ml. flask, equipped with a reflux condenser, a mixture of 1.00 g. (2.4 mmoles) of potassium tetrachloroplatinate(II) and 0.24 g. (5.6 mmoles) of lithium chloride in 30 ml. of tetrahydrofuran is stirred at 65° in an inert gas (argon). Subsequently, 0.8 ml. (9.8 mmoles) of allyl chloride (3-chloropropene) is added, followed by 0.48 g. (2.5 mmole) of tin(II) chloride. Upon the addition of the tin salt, the solution turns first deep red, then slowly yellow. After one hour at 65° the solution is filtered through a fritted-glass filter. The inorganic residue, a mixture of lithium chloride and unreacted potassium tetrachloroplatinate(II), weighs 0.57 g. The yellow solution is evaporated to dryness in a rotary evaporator. The residue is then recrystallized from 30 ml. of chloroform. The crystalline precipitate is filtered, washed with 15 ml. of ether, and dried in high vacuum. The crude product weighs 0.58 g. It is washed three times with water (total volume: 25 ml.) and three times with 5-ml. portions of acetone. After drying under vacuum, the yield is 0.57 g., which corresponds to 57% of theory based on K_2PtCl_4 used or 80% based on K_2PtCl_4 consumed.† *Anal.* Calcd. for $[C_3H_5ClPt]_4$: C, 13.3; H, 1.8; Cl, 12.9; Pt, 72.0. Found: C, 13.1; H, 1.7; Cl, 13.0; Pt, 73.0.

---

*The checkers used commercial anhydrous tin(II) chloride available from Alfa Inorganics, Box 159, Beverly, Mass. 01915.

†In a larger-scale reaction based on preformed Li_2PtCl_4 (obtained by passage of aqueous K_2PtCl_4 through a lithium-loaded sulfonic acid ion-exchange column), the editor obtained a 79% yield.

Properties

Tetrakis[allylchloroplatinum(II)] is an air-stable, yellow solid which is insoluble in organic solvents. It reacts with $Na[C_5H_5]$ to give $[(C_3H_5)Pt(C_5H_5)]$.[22] Triphenylphosphine converts it into (π-allyl)bis(triphenylphosphine)platinum(II) chloride, $[(\pi-C_3H_5)Pt(PPh_3)_2]Cl$. The tetramer has a cyclic structure as determined by G. Raper and W. S. McDonald.[23] Each platinum atom is bonded to two bridging chloride ligands as well as to two bridging allyl ligands.

References

1. J. Smidt and W. Hafner, *Angew. Chem.*, **71**, 284 (1959).
2. R. Huttel and J. Kratzer, *ibid.*, **71**, 456 (1959).
3. R. Huttel and H. Clerist, *Chem. Ber.*, **96**, 3101 (1963).
4. H. C. Volger, *Rec. Trav. Chim.*, **88**, 225 (1969).
5. J. Morelli, R. Ugo, F. Couti, and M. Donati, *Chem. Commun.*, **1967**, 801.
6. A. D. Ketley and J. Braatz, *ibid.*, **1968**, 169.
7. R. G. Schultz (Monsanto Chemical Co.), U.S. Patent 3,446,825 (1966).
8. M. Sakakibara, Y. Takakashi, S. Sakai, and Y. Ishii, *Chem. Commun.*, **1969**, 396.
9. B. L. Shaw, *Chem. Ind.*, **1962**, 1190.
10. S. D. Robinson and B. L. Shaw, *Proc. Chem. Soc.*, **1963**, 4806.
11. J. M. Rowe and D. A. White, *J. Chem. Soc. (A)*, **1967**, 1451.
12. J. Lukas, P. W. N. M. van Leeuwen, H. C. Volger, and P. Kramer, *Chem. Commun.*, **1970**, 799.
13. R. G. Schultz, *Tetrahedron*, **20**, 2809 (1964).
14. M. S. Lupin, J. Powell, and B. L. Shaw, *J. Chem. Soc. (A)*, **1966**, 1688.
15. J. Lukas, and J. E. Blom, *J. Organometallic Chem.*, **26**, C25 (1971).
16. A. I. Vogel, "A Textbook of Practical Organic Chemistry," Longmans, Green & Co., Ltd., London, 1951.
17. G. E. Coates, M. L. H. Green, and K. Wade, "Organometallic Compounds II," Methuen & Co., Ltd., London, 1968; M. I. Loback, B. D. Babitskii, and V. A. Kormer, *Russ. Chem. Rev.*, **36**, 476 (1967); R. Huttel, *Synthesis*, **1970**, 225.
18. J. R. Doyle, P. E. Slade, and H. B. Jonassen, *Inorganic Syntheses*, **6**, 218 (1960).
19. S. D. Robinson and B. L. Shaw, *J. Chem. Soc.*, **1963**, 4806; *ibid.*, **1964**, 5002.
20. J. M. Rowe and D. A. White, *J. Chem. Soc. (A)*, **1967**, 1451.
21. *Inorganic Syntheses*, **12**, 317 (1970).
22. W. Keim, dissertation, Technische Hochschule, Aachen, 1963.
23. G. Raper and W. S. McDonald, *Chem. Commun.*, **1970**, 655.

18. DICARBONYLCHLORO(*p*-TOLUIDINE)IRIDIUM(I)

$IrCl_3 \cdot 3H_2O + 3CO + LiCl \longrightarrow$

$$Li[Ir(CO)_2Cl_2] + 2HCl + CO_2 + 2H_2O$$

$Li[Ir(CO)_2Cl_2] + CH_3$—⟨O⟩—$NH_2 \longrightarrow$

$$IrCl(CO)_2(NH_2$$—⟨O⟩—$CH_3) + LiCl$

Submitted by U. KLABUNDE*
Checked by D. FORSTER† and W. O. WEINREIS†

Dicarbonylchloro(*p*-toluidine)iridium(I) has been prepared by the reaction of $K_2[Ir_2(CO)_2Cl_{4.8}]^1$ or $Ir(CO)_3Cl^2$ with *p*-toluidine. The preparation described here involves the carbonylation of iridium(III) chloride trihydrate in 2-methoxyethanol or ethanol in the presence of lithium chloride to give, presumably, the $[Ir(CO)_2$-$Cl_2]^-$ anion. (The intermediate has not been isolated but most likely is this anion,[3] previously isolated as its tetraphenylarsonium salt.[4]) Subsequent displacement of a chloride ion gives the title compound in high yield.

Procedure I

■ **Caution.** *This synthesis should be performed in a hood, and the pressure bottle should be adequately shielded. Carbon monoxide is a highly toxic, colorless, odorless gas!*

A 300-ml., heavy-walled, glass pressure-reaction bottle‡ containing a small magnetic stirring bar is charged with 3.5 g. of iridium-(III) chloride trihydrate, 1.0 g. of lithium chloride, and 50 ml. of 2-methoxyethanol. After attaching the pressure head, the bottle

*Central Research Department, E. I. du Pont de Nemours & Company, Wilmington, Del. 19898.
†Central Research Department, Monsanto Co., St. Louis, Mo. 63166.
‡A bottle for a Parr hydrogenation apparatus (A. H. Thomas Co., Philadelphia, Pa. 19105) is satisfactory, since the clamped rubber stopper fitting provides a seal adequate to hold 5 atmospheres pressure.

is evacuated, pressured with 40 p.s.i.g.* of carbon monoxide, and placed in a magnetically stirred oil bath. The bath is heated to 180° and kept at this temperature for 30 minutes; during this time the pressure increases to 55 p.s.i.g. and the color of the solution turns to a pale yellow. The oil bath is removed and the bottle allowed to cool to room temperature. After the pressure is released and the pressure head removed, a nitrogen hose is inserted into the bottle. (■ **Note.** *This solution is very oxygen-sensitive.*) To the stirred solution 1.3 g. of *p*-toluidine is added. The relatively air-stable solution is poured into 500 ml. of distilled water. The purple solid is collected on a filter, washed with 300 ml. of water, and dissolved in 400 ml. of benzene. (The checkers found it desirable to dry the crude solid and to filter the benzene solution, because a little insoluble residue was present.) After the solution is dried with anhydrous sodium sulfate, the benzene is evaporated under reduced pressure, and the residual dicarbonylchloro(*p*-toluidine)iridium(I) is dried *in vacuo*; yield is 98%. *Anal.* Calcd. for $C_9H_9ClIrNO_2$: C, 27.7; H, 2.32; N, 3.58. Found: C, 27.8; H, 2.16; N, 3.67.

Procedure II

A two-necked, 300-ml., round-bottomed flask equipped with magnetic stirring bar, gas inlet tube, and a condenser topped with a bubbler is charged with 3.5 g. of iridium(III) chloride trihydrate, 1.0 g. of lithium chloride, and 135 ml. of 2-methoxyethanol (or ethanol) and 15 ml. of water. Carbon monoxide is passed slowly through the solution via the gas inlet tube. The solution is stirred and boiled under reflux until it becomes pale yellow (usually 5 to 16 hours). After the solution has cooled to room temperature, the gas inlet tube is removed and 1.3 g. of *p*-toluidine is added to the stirred solution. The isolation of the product is identical to that described in Procedure I. The yield is 3.2 g. (75–90%).

Properties

Dicarbonylchloro(*p*-toluidine)iridium(I) is a purple, diamagnetic, crystalline solid that is readily soluble in polar organic solvents to

---

*Pressures are described in pounds per square inch gage (p.s.i.g.). Therefore, 40 p.s.i.g. = 55 p.s.i. absolute pressure = 3.7 atmospheres.

give slightly air-sensitive yellow to brown solutions. The solid decomposes above 160° and is stable to storage in air over a period of a year.

The ^1H n.m.r. spectrum in a 10% solution in acetone-d_6 shows the aromatic protons as a multiplet centered at $\tau 2.8$, the methyl protons at $\tau 7.60$, and the amino protons at $\tau 7.15$ in the area ratio of 4.0:3.0:2.3, respectively. The infrared spectrum (Nujol mull) shows ν_{CO} at 2088 and 2027 cm.$^{-1}$, and ν_{NH_2} at 3340 cm.$^{-1}$.

The complex serves as a useful intermediate for the preparation of other iridium(I) complexes, because the *p*-toluidine ligand can be readily displaced at room temperature by pyridine or 1,10-phenanthroline.[2] An additional carbon monoxide ligand is replaced with triphenylphosphine, -arsine, and -stibine as well as with phosphites and ethylenebis[diphenylphosphine].[2]

References

1. M. Angoletta, *Gazz. Chim. Ital.*, **89**, 2359 (1959).
2. W. Hieber and V. Frey, *Chem. Ber.*, **99**, 2607 (1966).
3. R. A. Schunn and W. G. Peet, *Inorganic Syntheses*, **13**, 126 (1972).
4. D. Forster, *Inorg. Nucl. Chem. Letters*, **5**, 433 (1969).

19. μ-NITRIDO-BIS(TRIPHENYLPHOSPHORUS)(1+) ("PPN") SALTS WITH METAL CARBONYL ANIONS

Submitted by J. K. RUFF* and W. J. SCHLIENTZ*
Checked by R. E. DESSY† and J. M. MALM† (Secs. A–C) and by G. R. DOBSON‡ and M. N. MEMERING‡ (Sec. D)

The isolation of labile complex anions is often facilitated by combination with bulky cations to give organic-soluble salts. These salts are often more stable and more easily crystallized than those with simple metallic cations. One of the most convenient cations has been the μ-nitrido-bis(triphenylphosphorus)(1+) ion, first isolated by Appel[1] as its bromide salt in a complicated four-step

*Department of Chemistry, University of Georgia, Athens, Ga. 30601.
†Department of Chemistry, Virginia Polytechnic Institute, Blacksburg, Va. 24061.
‡Department of Chemistry, North Texas State University, Denton, Tex. 76203.

synthesis. Since this cation was found to be an especially convenient counterion for the isolation of crystalline salts of mononuclear and polynuclear carbonyl anions,[2-4] a more simple synthesis was developed. This synthesis (Sec. A), somewhat analogous to one reported by Becke-Goehring,[5] is accomplished with readily available starting materials.

μ-Nitrido-bis(triphenylphosphorus)(1+) chloride (trivial designation [PPN]Cl is readily metathesized with alkali-metal salts of other simple inorganic anions, as illustrated for nitrate in Sec. B. Bromide, iodide, cyanide, cyanate, thiocyanate, azide, and nitrite salts have been prepared similarly.[6]

The use of PPN cation to precipitate the very labile tetracarbonyl-cobaltate(1−) anion is described in Sec. C. Not only is the salt soluble in many polar organic solvents, but it possesses surprising stability in the solid state. Similar properties have been noted for the PPN salts of $[Mn(CO)_5]^-$ and $[V(CO)_6]^-$.

The properties of the PPN cation have also facilitated the crystallization of salts of the dinuclear metal carbonyl anions, $[M_2(CO)_{10}]^{2-}$ (M = Cr, Mo, W), and permitted a crystal-structure determination[7] of these remarkable complexes. The dinuclear chromium decacarbonyl anion was first prepared by Behrens and Vogl[8] by the reaction of $Cr(CO)_6$ with $NaBH_4$ in liquid NH_3 at 40°. More recently Hayter[9] prepared dinuclear decacarbonyl anions of all three Group VI metals by a photochemical process and isolated them as the tetraethylammonium salts. A convenient synthesis involving isolation of the PPN salts of these anions is described in Sec. D.

A. μ-NITRIDO-BIS(TRIPHENYLPHOSPHORUS)(1+) CHLORIDE

$$2(C_6H_5)_3P + 2Cl_2 \longrightarrow 2(C_6H_5)_3PCl_2$$

$$2(C_6H_5)_3PCl_2 + (C_6H_5)_3P + NH_2OH \cdot HCl \longrightarrow$$

$$[\{(C_6H_5)_3P\}_2N]Cl + (C_6H_5)_3PO + 4HCl$$

Procedure

■ **Caution.** *This reaction should be carried out in an efficient fume hood because of the hydrogen chloride liberated.*

The synthesis is carried out in a 2-1. three-necked flask equipped with mechanical stirrer (the mixture becomes very viscous), gas inlet tube, and a reflux condenser fitted with a drying tube or nitrogen bubbler to exclude moisture. One liter of 1,1,2,2-tetrachloroethane and 786 g. of triphenylphosphine* are stirred until the latter dissolves. The solution is cooled to a temperature between -20 and $-30°$ by use of a cold bath prepared from Dry Ice and 2-propanol. A lecture bottle of chlorine, attached to the gas inlet tube by Tygon tubing, is placed on a balance and weighed. Chlorine is slowly added to the solution until a weight loss of 142 g. has occurred. (After *ca.* 40 g. has been added, separation of a solid impedes stirring and heat transfer.) The gas inlet tube is replaced with a stopper. Hydroxylamine hydrochloride (69 g.) is added to the solution, and the mixture is allowed to warm to room temperature. A heating mantle is placed around the flask, and the mixture is boiled under reflux until hydrogen chloride evolution practically ceases (usually 6 to 8 hours). After the solution has cooled to room temperature, it is poured into 4 l. of ethyl acetate. The product crystallizes upon standing overnight. The crude μ-nitrido-bis[triphenylphosphorus]-(1+) chloride is recrystallized from boiling water; approximately 1 l. of water per 100 g. of salt is used. The recrystallized product is dried under vacuum (10^{-3} torr) overnight; yield, 538 g. (92%); m.p. 268–270°. *Anal.* Calcd. for $C_{36}H_{30}ClN_2P$: C, 75.5; H, 5.26; N, 2.44; Cl, 6.18. Found: C, 75.3; H, 5.15; N, 2.38; Cl, 6.05.

Properties

μ-Nitrido-bis(triphenylphosphorus)(1+) chloride is an air-stable, nonhygroscopic compound which may be recrystallized from boiling water without decomposition. It may also be recrystallized from a dichloromethane–diethyl ether mixture. In this case, however, the product is obtained as a dichloromethane solvate (m.p. 260–262°). Heating at 60° under a vacuum of 10^{-3} torr converts it to the unsolvated salt. Other simple inorganic salts behave in a similar manner. The inorganic μ-nitrido-bis(triphenylphosphorus)(1+) salts are sol-

---

*The scale of this reaction may be reduced substantially, but the yield is adversely affected because the accuracy with which the chlorine can be measured decreases. The checkers obtained 65–85% yields working at one-tenth this scale.

uble in ethanol, dichloromethane, and acetonitrile but are insoluble in tetrahydrofuran, diethyl ether, and nonpolar organic solvents.

The ^{31}P n.m.r. spectrum of μ-nitrido-bis(triphenylphosphorus)-$(1+)$ chloride consists of a singlet at -22.3 p.p.m. vs. 85% phosphoric acid,[6] indicating that both phosphorus atoms are equivalent. The crystal structure[7] of the $[Cr_2(CO)_{10}]^{2-}$ salt confirms the equivalency of the phosphorus atoms, since both P—N distances are equal (1.58 A.).

B. μ-NITRIDO-BIS(TRIPHENYLPHOSPHORUS)(1+) NITRATE

Procedure

A 25.0-g. sample of μ-nitrido-bis(triphenylphosphorus)(1+) chloride is dissolved in 500 ml. of boiling water. This solution is added to a boiling solution of 250 g. of potassium nitrate in 200 ml. of water. Precipitation is immediate, and the mixture is filtered while still warm. The precipitate is washed several times with cold water and is recrystallized from 1500 ml. of boiling water. The nitrate salt, obtained in 24-g. yield (92%, m.p. 224–225°), contains a trace of chloride which can be removed by repeating the above treatment. *Anal.* Calcd. for $C_{36}H_{30}N_2O_3P_2$: C, 72.0; H, 5.03; N, 4.66. Found: C, 71.8; H, 4.97; N, 4.69.

C. μ-NITRIDO-BIS(TRIPHENYLPHOSPHORUS)(1+) TETRACARBONYLCOBALTATE(1−)

$$Co_2(CO)_8 + 2Na(Hg) \longrightarrow 2Na[Co(CO)_4]$$

$$Na[Co(CO)_4] + [\{(C_6H_5)_3P\}_2N]Cl \longrightarrow$$

$$[\{(C_6H_5)_3P\}_2N][Co(CO)_4] + NaCl$$

Procedure

■ **Caution.** *Cobalt carbonyl and its salts are toxic.*

Octacarbonyldicobalt (3.42 g., 10 mmoles) is weighed out in a nitrogen-filled plastic glove bag and transferred to a 250-ml. single-necked flask containing 50 g. 1% sodium amalgam in 100 ml. of anhydrous tetrahydrofuran. The mixture is stirred for 3 hours under a nitrogen atmosphere. A solution of 10.0 g. (17.4 mmoles*) of μ-nitrido-bis(triphenylphosphorus)(1+) chloride in 100 ml. of dichloromethane is added to the mixture. After stirring for 10 minutes, the supernatant liquid is decanted from the excess amalgam into another flask. Celite filter aid is added, and the liquid is filtered.† The solvent is removed from the filtrate under reduced pressure. The residue is redissolved in 100 ml. of dichloromethane. (Filtration is repeated if a clear solution is not obtained.) Diethyl ether (150–200 ml.) is slowly added to the solution until precipitation of the product begins. The flask is flushed with nitrogen, closed, and placed in a freezer. After precipitation appears to be complete, the mixture is filtered, and the solid $[\{(C_6H_5)_3P\}_2N][Co(CO)_4]$ is dried under vacuum at 50°; yield is 9.71 g. (78.5%); decomposition point is 160–165°. *Anal.* Calcd. for $C_{40}H_{30}CoNO_4P_2$: C, 68.7; H, 4.29; Co, 8.32; N, 2.00. Found: C, 68.3; H, 4.39; Co, 8.02; N, 1.95.

Properties

μ-Nitrido-bis(triphenylphosphorus)(1+) tetracarbonylcobaltate-(1−) is stable in the air for prolonged periods of time. It is more soluble in organic solvents than is the chloride. For example, it is soluble in nitromethane and tetrahydrofuran. A single band in the carbonyl stretching region of the infrared spectrum occurs at 1890 cm.$^{-1}$.

D. μ-NITRIDO-BIS(TRIPHENYLPHOSPHORUS)(1+) SALTS OF $[M_2(CO)_{10}]^{2-}$ (M = Cr, Mo, W)

$$2M(CO)_6 + 2Na(Hg) \longrightarrow Na_2[M_2(CO)_{10}] + 2CO$$

$$Na_2[M_2(CO)_{10}] + 2[\{(C_6H_5)_3P\}_2N]Cl \longrightarrow$$
$$\text{"PPN Cl"}$$

$$[\{(C_6H_5)_3P\}_2N]_2[M_2(CO)_{10}] + 2NaCl$$

---

*Less than a stoichiometric quantity of the phosphorus salt is used because formation of $Na[Co(CO)_4]$ is not quantitative.

†Although most carbonyl anions are sensitive to oxidation by atmospheric oxygen, it is not necessary to use an inert atmosphere after the addition of the μ-nitrido-bis(triphenylphosphorus)(1+) chloride has been completed if prolonged exposure to the atmosphere is avoided.

Procedure

■ **Caution.** *Group VI metal carbonyls are volatile and toxic. Hence they should be handled in an efficient fume hood.*

To a 500-ml. round-bottomed flask containing 300 ml. of tetrahydrofuran (freshly distilled from lithium tetrahydridoaluminate)[10] is added 100 g. of 0.75% sodium amalgam and 20 mmoles of metal hexacarbonyl [7 g. $W(CO)_6$, 5.2 g. $Mo(CO)_6$, or 4.4 g. $Cr(CO)_6$]. The flask is fitted with a reflux condenser with nitrogen bubbler. The mixture is irradiated 14 to 18 hours with a G.E. H44-4GS-type mercury-vapor spotlight or equivalent* with vigorous stirring. An egg-shaped stirring bar and a heavy-duty magnetic stirrer are used. The solution is decanted from mercury into a 500-ml. round-bottomed flask containing a solution of 6 g. (10.5 mmoles†) of recrystallized μ-nitrido-bis(triphenylphosphorus)(1+) in 50 ml. of dichloromethane. The solution is evaporated to dryness *in vacuo*. The residue is dissolved with 50 ml. of dichloromethane and filtered on a coarse-porosity sintered-glass filter with Celite filter aid. After purging with nitrogen, 75–100 ml. of ethyl acetate is slowly added to the mixture, which is allowed to stand at $-10°$. The precipitated salt is filtered and washed with 50 ml. of ethanol to remove traces of chloride. It is then washed with 50 ml. of pentane and is dried under high vacuum (10^{-4} torr). The compounds are reasonably stable for several months when stored under nitrogen in the dark in a freezer.

Properties

All three products are soluble in acetone and dichloromethane, sparingly soluble in tetrahydrofuran, and insoluble in diethyl ether, ethyl acetate, ethanol, and pentane.

---

*The source of radiation is relatively unimportant as long as Pyrex is used as the container. A G.E. H44-4GS lamp or G.E. H100 PSP 44-4 lamp is suitable. A medium-pressure 450-watt Hanovia lamp was employed by the checkers. The lamp used by the authors was obtained from E. Sam Jones Co., Atlanta, Ga.

†Less than a stoichiometric quantity of phosphorus salt is used, because formation of $Na_2[M_2(CO)_{10}]$ is not quantitative. Increasing the amount of phosphorus salt does not increase the yield.

Bis [μ-nitrido-bis[triphenylphosphorus]$(1+)$] [decacarbonylditungstate]$(2-)$ is a bright yellow-orange crystalline compound, m.p. 205–207° (decomposes). The yields range from 3 to 8.5 g. (20–50%), with 5 g. being a typical yield. Analysis of the compound as the dichloromethane solvate is: Calcd. C, 55.1; H, 3.5; N, 1.5. Found: C, 55.3; H, 3.7; N, 1.7.

Bis [μ-nitrido-bis[triphenylphosphorus]$(1+)$] [decacarbonyldimolybdate]$(2-)$ is a rust-colored crystalline compound, m.p. 193–195° (decomposes). The yields range from 2 to 4 g. (15–25%), with 3 g. being a typical yield. Analysis of the compound as the dichloromethane solvate is: Calcd.: C, 61.0; H, 3.8; N, 1.7; Cl, 4.3. Found: C, 60.9; H, 3.7; N, 1.8; Cl, 4.6.

Bis [μ-nitrido-bis[triphenylphosphorus]$(1+)$] [decacarbonyldichromate]$(2-)$ is a dark yellow to red-brown crystalline compound, m.p. 185–189° (decomposes). The yields range from 3 to 5.5 g. (20–40%), with 4 g. being a typical yield. Analysis of the dichloromethane solvate is: Calcd.: C, 64.5; H, 4.0; N, 1.8; Cl, 4.6. Found: C, 64.1; H, 4.2; N, 1.9; Cl, 5.0.

References

1. R. Appel and A. Hauss, *Z. Anorg. Allgem. Chem.*, **311**, 290 (1961).
2. J. K. Ruff, *Inorg. Chem.*, **7**, 1818 (1968).
3. J. K. Ruff, *ibid.*, **7**, 1821 (1968).
4. J. K. Ruff, *ibid.*, **7**, 1499 (1968).
5. M. Becke-Goehring and E. Fluch, *Inorganic Syntheses*, **8**, 92 (1966).
6. J. K. Ruff, unpublished results.
7. L. B. Handy, J. K. Ruff, and L. F. Dahl, *J. Am. Chem. Soc.*, **92**, 7312 (1970).
8. H. Behrens and J. Vogl, *Chem. Ber.*, **96**, 2220 (1963).
9. R. G. Hayter, *J. Am. Chem. Soc.*, **88**, 4376 (1966).
10. *Inorganic Syntheses*, **12**, 317 (1970).

20. DICARBONYL-h^5-CYCLOPENTADIENYLNITROSYL-MANGANESE(1+) HEXAFLUOROPHOSPHATE(1−)

$$[Mn(C_5H_5)(CO)_3] + NOPF_6 \longrightarrow [Mn(C_5H_5)(CO)_2(NO)][PF_6] + CO$$

Submitted by N. G. CONNELLY*
Checked by R. B. KING† and Y. WATANABE†

Dicarbonyl-h^5-cyclopentadienylnitrosylmanganese(1+) hexafluoro-phosphate(1−) is usually prepared[1] by the reaction of tricarbonyl-h^5-cyclopentadienylmanganese with sodium nitrite and hydrochloric acid in refluxing ethanol. The cationic nitrosyl is isolated by the addition of ammonium hexafluorophosphate to the reaction mixture. This procedure yields up to 50% of the product in a total reaction time of several hours. A more efficient preparation is the direct reaction of $[Mn(C_5H_5)(CO)_3]$ with $[NO][PF_6]$‡ in acetonitrile, which allows the isolation of the cationic complex in quantitative yield in less than one hour.

Procedure

■ **Caution.** *Because of the toxicity of manganese carbonyl derivatives, the preparation must be carried out in a well-ventilated hood. Acetonitrile is also toxic.*

The reaction is conducted in an atmosphere of dry nitrogen, and the acetonitrile must be dried before use (molecular sieves are satisfactory), since nitrosyl hexafluorophosphate is easily hydrolyzed. To 2 g. of tricarbonyl-h^5-cyclopentadienylmanganese[1] in 100 ml. of acetonitrile is added dropwise, and with rapid stirring, a solution of 1.9 g. $[NO][PF_6]$ in 30 ml. of acetonitrile. Carbon monoxide is evolved, and the solution becomes darker yellow. After the addition of the $[NO][PF_6]$ solution is complete (5 minutes), stirring is continued for 10 minutes to ensure complete reaction. The volume of the solvent is reduced to approximately 20 ml. by evaporation,

*School of Chemistry, University of Bristol, Bristol, England.
†Department of Chemistry, University of Georgia, Athens, Ga. 30601.
‡Ozark Mahoning Co., 1870 South Boulder, Tulsa, Okla. 74119.

and ether is added until precipitation of the product is complete (approximately 50 ml.). The yellow solid is filtered, recrystallized from acetone-ether, and air-dried to give dicarbonyl-h^5-cyclopentadienylnitrosylmanganese(1+) hexafluorophosphate(1−) as a yellow solid in 90–95% yield. The infrared spectrum (KBr disk) showed ν_{CO} at 2125 and 2075 cm.$^{-1}$, ν_{NO} at 1840 cm.$^{-1}$, and strong P—F absorption at 835 cm.$^{-1}$. *Anal.* Calcd. for $C_7H_5F_6MnNO_3P$: C, 23.96; H, 1.44; N, 3.99. Found: C, 23.84; H, 1.29; N, 4.14.

Properties

Dicarbonyl-h^5-cyclopentadienylnitrosylmanganese(1+) hexafluorophosphate(1 −) is an air-stable solid with no well-defined melting point. The ^{1}H n.m.r. spectrum (acetone solution) is a sharp singlet at $\tau 3.86$.[1] The complex reacts with donor ligands (L) such as phosphines, pyridines, isocyanides, and phosphites, and bidentate ligands (L—L) such as 2,2′-bipyridyl and 1,10-phenanthroline to give complexes of the type $[Mn(C_5H_5)(CO)(NO)L][PF_6]$, $[Mn(C_5H_5)(NO)L_2][PF_6]$, and $[Mn(C_5H_5)(NO)(L—L)][PF_6]$.[2-4] The complex $[Mn(C_5H_5)(CO)(NO)(PPh_3)][PF_6]$ has been resolved into optical isomers.[5] Bidentate sulfur ligands react to give complexes of the type $[(C_5H_5)(NO)(S—S)]^z$, where S—S is $[S_2C_2(CN)_2]^{-2}$, toluene-3, 4-dithiol, tetrachlorobenzenedithiol or $[S_2CN(R)_2]^{-1}$, and $z = 0$ or $- 1$.[6] Nucleophilic attack by $OCH_3{}^-$ on one of the carbonyl ligands to give $[Mn(CO_2CH_3)(C_5H_5)(CO)(NO)]$ has also been reported.[7] Reduction by sodium tetrahydroborate results in the formation of $[Mn(C_5H_5)(CO)_2(NO)]_2$.[8]

References

1. "Organometallic Syntheses," R. B. King (ed.), Vol. 1, Academic Press, Inc., New York, 1965.
2. H. Brunner and H.-D. Schindler, *J. Organometallic Chem.*, **19**, 135 (1969).
3. R. B. King and A. Efraty, *Inorg. Chem.*, **8**, 2374 (1969).
4. T. A. James and J. A. McCleverty, *J. Chem. Soc. (A)*, **1970**, 850.
5. H. Brunner and H.-D. Schindler, *J. Organometallic Chem.*, **24**, C7 (1970).
6. J. A. McCleverty, T. A. James, and E. J. Wharton, *Inorg. Chem.*, **8**, 1340 (1969).
7. R. B. King, M. B. Bisnette, and A. Fronzaglia, *J. Organometallic Chem.*, **4**, 256 (1965).
8. R. B. King and M. B. Bisnette, *J. Am. Chem. Soc.*, **85**, 2527 (1963).

21. HEXAAMMINEPLATINUM(IV) CHLORIDE

Submitted by L. N. ESSEN*
Checked by F. S. WAGNER† and J. K. BEATTIE†

Several methods for the synthesis of hexammineplatinum(IV) chloride have been described, but the best procedure[1] involves the prior preparation of $[Pt(NH_2CH_3)_4Cl_2]Cl_2$. This complex can be converted first to hexaammineplatinum(IV) sulfate, which in turn is converted to the corresponding chloride. The procedure itself is very simple and safe, and the complex obtained is relatively pure, which represents the advantage over the other published methods.[2] The checkers' analyses indicate retention of *ca.* 1% of the methylamine ligands in the final product. The synthesis of $[Pt(NH_3)_6]Cl_4$ from $K_2[PtCl_6]$, and from $K_2[PtCl_4]$, using $[Pt(NH_2CH_3)_4Cl_2]Cl_2$ as the intermediary product, is described.

A. DICHLOROTETRAKIS(METHYLAMINE)PLATINUM(IV) CHLORIDE

1. From $K_2[PtCl_6]$

$$K_2[PtCl_6] + 4 CH_3NH_2 \longrightarrow [Pt(NH_2CH_3)_4Cl_2]Cl_2 + 2 KCl$$

Procedure

■ **Caution.** *The chlorine addition should be carried out in an efficient fume hood.*

To 3 g. of $K_2[PtCl_6]$, 30 ml. of water and 20 ml. of a 30% methylamine solution are added. The mixture is heated on a hot plate for 30–40 minutes until the solid dissolves to give a yellow solution. After filtration, the filtrate is evaporated to 20–30 ml. on a steam bath. To the solution so obtained, 30 ml. of warm 12 *M* hydrochloric acid is added, and the heating on the steam bath is continued for another 40–50 minutes while a stream of chlorine is passed through the solution. After cooling the solution, the precipitated yellow

*Akademiya Nauk SSSR, Leninski Prospekt 31, Moskva V-71, SSSR.
†Department of Chemistry, University of Illinois, Urbana, Ill. 61801.

product is filtered off and washed with ethanol and ether. The yield amounts to about 40%. The product obtained can be used for the hexaammine synthesis without further purification. *Anal.* Calcd. for $C_4H_{20}Cl_4N_4Pt$: C, 10.41; H, 4.37; N, 12.15; Pt, 42.31. Found (checkers' results): C, 9.47; H, 4.18; N, 11.87; Pt, 43.10.

2. From $K_2[PtCl_4]^*$

$$K_2[PtCl_4] + Cl_2 + 4\,CH_3NH_2 \longrightarrow [Pt(NH_2CH_3)_4Cl_2]Cl_2 + 2\,KCl$$

Procedure

■ **Caution.** *The chlorine addition should be carried out in an efficient fume hood.*

To 60 ml. of water is added 1.8 g. of $K_2[PtCl_4]^3$ and 5 ml. of 40% methylamine solution. The solution is warmed on a steam bath for 15 minutes, during which time the color of the solution changes from dark red to light yellow. (Heating on a hot plate results in precipitation of the sparingly soluble Magnus-type salt,[3] which redissolves very slowly.) The solution is evaporated to dryness with a rotary evaporator to ensure complete removal of excess methylamine. The residue is dissolved in 30 ml. of 6 *M* HCl, and chlorine gas is bubbled through the solution for 30 minutes. Upon completion of the chlorine oxidation, the yellow precipitate which forms is collected by filtration and is washed with cold water, absolute ethanol, and diethyl ether. Yield is 1.6 g. (82%). *Anal.* Calcd. for $C_4H_{20}Cl_4N_4Pt$: C, 10.41; H, 4.37; N, 12.15; Pt, 42.31. Found: C, 10.34; H, 4.31; N, 12.09; Pt, 42.46.

B. HEXAAMMINEPLATINUM(IV) CHLORIDE

$$[Pt(NH_2CH_3)_4Cl_2]Cl_2 + 2\,(NH_4)_2SO_4 + 2\,NH_3 \longrightarrow$$
$$[Pt(NH_3)_6](SO_4)_2 + 4\,[CH_3NH_3]Cl$$
$$2[Pt(NH_3)_6][SO_4]_2 + 2\,NaOH \longrightarrow$$
$$[Pt(NH_3)_5NH_2]_2[SO_4]_3 + Na_2SO_4 + 2H_2O$$
$$[Pt(NH_3)_5NH_2]_2[SO_4]_3 + Na_2SO_4 + 4\,BaCl_2 \longrightarrow$$
$$4\,BaSO_4 + 2\,NaCl + 2\,[Pt(NH_3)_5NH_2]Cl_3$$
$$[Pt(NH_3)_5NH_2]Cl_3 + HCl \longrightarrow [Pt(NH_3)_6]Cl_4$$

*Contributed by F. S. Wagner and J. K. Beattie.

Procedure

To 1 g. of $[Pt(NH_2CH_3)_4Cl_2]Cl_2$ is added a solution of 4 g. of $(NH_4)_2SO_4$ in 15 ml. of water and 10 ml. of 15 M ammonia solution, and the mixture so obtained is boiled on a hot plate for 30–40 minutes. During this time, a white precipitate is formed. It is useful to repeat this boiling twice, adding, every 10 minutes, 7–10 ml. of concentrated ammonia.

After 1.5 hours of standing, the precipitate of hexaammineplatinum-(IV) sulfate is filtered, washed once or twice with small amounts of ethanol and ether, and dried at 60–80°. Yield is about 50%.

In order to convert the hexaammineplatinum(IV) sulfate to hexaammineplatinum(IV) chloride, the weighed amount of hexa-ammine sulfate is treated with a small amount of water and then with 30% NaOH solution until the precipitate is completely dissolved (the NaOH solution should be added dropwise in order to avoid an excess). The hexaammine sulfate (approximately 40% $SO_4{}^{2-}$) is treated with the calculated amount of $BaCl_2 \cdot 2H_2O$, which should be weighed out carefully to give only a very small excess. The $BaCl_2 \cdot 2H_2O$ is dissolved in a small amount of boiling water and added dropwise to the boiling solution of the hexaammine sulfate.

The precipitated $BaSO_4$ is filtered off after 2 hours, and the filtrate obtained is acidified strongly with concentrated hydrochloric acid. A white precipitate of hexaammineplatinum(IV) chloride is formed immediately. It may be necessary to reduce the volume of the solution by evaporation in order to precipitate the product. It is filtered and washed with ethanol. The yield amounts to about 80% based on $[Pt(NH_3)_6](SO_4)_2$ or to about 16% based on $K_2[PtCl_6]$. The following analyses were supplied by the checkers: Calcd. for $[Pt(NH_3)_6]Cl_4 \cdot H_2O$: Pt, 42.67; H, 4.41; N, 18.39; Cl, 31.02. Found using intermediate prepared from $K_2[PtCl_4]$: Pt, 42.34; H, 4.39; N, 17.93; Cl, 30.76; C, 0.24. Found using intermediate prepared from $K_2[PtCl_6]$: Pt, 42.71; H, 4.43; N, 17.92; Cl, 30.32; C, 0.31.

Properties

Hexaammineplatinum(IV) chloride forms snow-white, triclinic crystals with refractive indices: $n_\alpha = 1.717$; $n_\gamma = 1.724$. Its solubility in water is 2.01 g./100 g. at 0°.

References

1. Kh. I. Gildengershel, Zhur. Priklad. Khim., 23 (5), 487 (1950).
2. "Synthesis of Platinum Metals Group Complex Compounds," I. I. Chernyaev (ed.), p. 167, Nauka.
3. G. B. Kauffman and D. O. Cowan, Inorganic Syntheses, 7, 239 (1963).

22. BIS(1,1,1,5,5,5-HEXAFLUORO-2,4-PENTANEDIONATO)-NICKEL(II) AND -COBALT(II)

(Nickel and Cobalt Hexafluoroacetylacetonates)

Submitted by R. L. PECSOK* and W. D. REYNOLDS† (Method A) and by J. P. FACKLER, JR.,‡
I. LIN,‡ and J. PRADILLA-SORZANO‡ (Method B)
Checked by J. P. FACKLER, JR.,‡ and I. LIN‡ (Method A) and by G. W. PARSHALL§ and
F. R. WONCHOBA (Method B)

Bis- and tris(hexafluoro-2,4-pentanedionato) (hexafluoroacetyl-acetonate, hfa) complexes of 23 different metal(II), (III), and (IV) ions have been prepared.[1-5] Most of these have been isolated as low-melting solids, which can be readily sublimed unchanged between 25 and 150°. Most methods of synthesis of this class of complexes in different solvents have resulted in incomplete conversion to the chelate. For example, the copper complex $Cu(hfa)_2$ has been prepared in 80% yield using excess ligand as the solvent. Several solvents having various polarities (water, ethanol, carbon tetrachloride, acetone) have been used.

Two convenient methods of synthesis of the nickel(II) and cobalt-(II) complexes are described here. Method A, based on reaction of the sodium enolate salt of 1,1,1,5,5,5-hexafluoro-2,4-pentanedione with the transition-metal chlorides in dimethylformamide (dmf), gives almost quantitative yields of $Ni(hfa)_2(dmf)_2$ and $Co(hfa)_2(dmf)_2$. Method B gives lower yields (50–75%) of the same products but can be carried out rapidly in common laboratory equipment (Method A is best carried out by vacuum-line techniques[7]).

*Department of Chemistry, University of Hawaii, 2545 The Mall, Honolulu, Hawaii 96822.
†Vallecitos Nuclear Center, General Electric Co., Pleasanton, Calif.
‡Department of Chemistry, Case Western Reserve University, Cleveland, Ohio 44106.
§Central Research Department, E. I. du Pont de Nemours & Company, Wilmington, Del. 19898.

METHOD A

$$2Na + 2CF_3COCH_2COCF_3 \longrightarrow 2Na[CF_3COCHCOCF_3] + H_2$$

$$MCl_2 + 2Na[CF_3COCHCOCF_3] \xrightarrow{\text{HCONMe}_2}$$

$$[M(CF_3COCHCOCF_3)_2 \cdot (HCONMe_2)_2] + 2NaCl$$

$$M = Ni, Co$$

Procedure

■ **Caution.** *Hexafluoro-2,4-pentanedione is a potentially hazardous, volatile liquid. It should always be handled in an efficient fume hood.*

1,1,1,5,5,5-Hexafluoro-2,4-pentanedione* is purified on a conventional high-vacuum line[7] by distillation from P_2O_5 through a 30-cm. 8-mm.-diam. low-temperature column at $-55°$ (*ca.* 2 torr). The purity may be checked by titration in $\simeq 15\%$ ethanol in distilled water and by gas-liquid chromatography. Reagent-grade dimethylformamide is distilled from anhydrous $MgSO_4$ and stored under dry nitrogen. Spectro-grade carbon tetrachloride, chloroform, and nitroethane solvents are used without further purification. The pyridine is reagent-grade, distilled from anhydrous BaO, and stored under anhydrous conditions.

1. Preparation of the Sodium Enolate of Hexafluoro-2,4-pentanedione

Freshly cut sodium (*ca.* 2 g.) is weighed and transferred to a vacuum-line reaction flask under anhydrous conditions and allowed to react with an excess of the ligand for 2 hours at 0°. The mixture is warmed to room temperature and stirred for an additional 24 hours. The excess ligand is removed on the high-vacuum line, and the sodium enolate is stirred under dry nitrogen.

2. Bis(dimethylformamide)bis(1,1,1,5,5,5-hexafluoro-2,4-pentanedionato)nickel(II).

About 0.25 g. (0.002 mole) of anhydrous nickel(II) chloride (dried at 110° for 3 days) is dissolved in 50 ml. of dimethylformamide. A small excess (2.3 g., 0.010 mole) of the sodium enolate prepared above is also dissolved in dimethylformamide. The two solutions

---

*Available from Peninsular ChemResearch, Box 14318, Gainesville, Fla. 32603.

are mixed under anhydrous conditions and refluxed at 100° for at least 2 hours. During the reaction period the color changes from yellow to green, after which time the excess solvent is removed in a rotary vacuum evaporator. The remaining residue is extracted with chloroform, filtered, washed with distilled water, and dried. The green solid is recrystallized from chloroform and dried in a desiccator under vacuum; yield is 1.25 g. (98–99%); m.p. 111–112°. *Anal.* Calcd. for $C_{16}H_{16}F_{12}N_2NiO_6$: C, 31.02; H, 2.60; N, 4.52; Ni, 9.48; mol. wt., 619. Found: C, 31.03; H, 2.48; N, 4.55; Ni, 9.36; mol. wt., 633 ± 32 (by thermoelectric vapor-pressure osmometer on 0.01 to 0.03 M solutions in benzene).

3. Bis(dimethylformamide)bis(1, 1, 1, 5, 5, 5-hexafluoro-2, 4-pentanedionato)cobalt(II)

The procedure is identical to that for the nickel chelate, using anhydrous cobalt(II) chloride as a starting material. The product forms tan, needle-shaped crystals; yield is 1.20 g. (97–98%); m.p. 89–90°. *Anal.* Calcd. for $C_{16}H_{16}F_{12}N_2O_6Co$: C, 31.03; H, 2.61; N, 4.52; mol. wt., 619. Found: C, 31.01; H, 2.75; N, 4.51; mol. wt., 653 ± 32 (by thermoelectric vapor-pressure osmometer on 0.01 to 0.03 M solutions in benzene).

METHOD B

$$MCl_2 \cdot 6H_2O + 2CF_3COCH_2COCF_3 + 2Na[O_2CCH_3] \xrightarrow{\text{HCONMe}_2}$$

$$[M(CF_3COCHCOCH_3)_2 \cdot (HCONMe_2)_2] + 2CH_3COOH + 2NaCl$$

$$M = Ni, Co$$

Procedure

1. Bis(dimethylformamide)bis(1,1,1,5,5,5-hexafluoro-2,4-pentanedionato)cobalt(II)

To 4.76 g. (0.02 mole) of $CoCl_2 \cdot 6H_2O$ dissolved in 90 ml. of dimethylformamide, 8.32 g. (0.04 mole) of 1,1,1,5,5,5-hexafluoro-2,4-pentanedione is added (■ **Caution.** *Handle in hood*). A solution of 3.6 g. (*ca.* 0.04 mole) of sodium acetate in 75 ml. of distilled water is added. Some solid precipitates immediately from the clear, red solution. The mixture is stirred for 2 hours (evaporation of hfa can be reduced by using a lightly stoppered container) and poured into a 500-ml. beaker containing some crushed ice. After stirring an additional 5 minutes, the solid is filtered, washed with water, and dried under vacuum. The crude complex is dissolved in a minimum amount (25 ml.) of dimethylformamide, and the solvent is removed by evaporation on a rotary vacuum evaporator or with flowing nitrogen gas. The product is crystallized from 40–50 ml. of chloroform to give 9.43 g. (76% yield) of material identical to $Co(hfa)_2(dmf)_2$ prepared by Method A in melting point (89–90°) and infrared spectrum. *Anal.* Found: C, 31.10; H, 2.56; N, 4.35.

2. Bis(dimethylformamide)bis(1,1,1,5,5,5-hexafluoro-2,4-pentanedionato)nickel(II)

Using a procedure similar to that for cobalt, 2.38 g. of $NiCl_2 \cdot 6H_2O$ in 200 ml. of dimethylformamide gives $Ni(hfa)_2(dmf)_2$ in 50–75% yield after recrystallization. The melting point, mixed melting point, and infrared spectrum are identical to those of the product of Method A. *Anal.* Found: C, 31.07; H, 2.45.

Properties[3]

The 1,1,1,5,5,5-hexafluoro-2,4-pentanedionato complexes of nickel-(II) and cobalt(II), when isolated as dimethylformamide adducts, are crystalline solids, stable to at least 135° *in vacuo.* The crystalline nickel complex is unaffected by immersion in water at room temperature for 3 days.

The infrared spectrum of $Ni(hfa)_2(dmf)_2$ in a KBr pellet shows three strong bands at 1655, 1648, and 1640 cm.$^{-1}$. The 1655 cm.$^{-1}$ absorption is tentatively assigned to the $\mathrm{>C\cdots O}$ stretching frequency of coordinated dimethylformamide. This frequency occurs at 1665 cm.$^{-1}$ in the free ligand and presumably would shift to the lower frequency upon adduct formation. The band at 1648 cm.$^{-1}$ is assigned to the $\mathrm{>C\cdots O}$ stretching mode from the hfa chelate.[8] The remaining band at 1640 cm.$^{-1}$ is assigned to a $\mathrm{>C\cdots C<}$ stretching mode.

Both the cobalt and the nickel complexes presumably have an octahedral coordination geometry, as indicated by their optical spectra.[9]

References

1. R. N. Hazeldine, W. K. R. Musgrave, F. Smith, and L. M. Turton, *J. Chem. Soc.,* **1951,** 609.
2. R. W. Moshier and R. E. Sievers, "Gas Chromatography of Metal Chelates," p. 139, Pergamon Press, Ltd., Oxford, 1965.
3. W. D. Reynolds, *Diss. Abstr.,* **26,** 3010 (1965).
4. S. C. Chattoraj and R. E. Sievers, *Inorg. Chem.* **6,** 408 (1967).
5. H. Veening, W. E. Bachmann, and D. M. Wilkinson, *J. Gas Chromatog.,* **5,** 248 (1967).
6. R. L. Belford, A. E. Martell, and M. Calvin, *J. Inorg. Nucl. Chem.,* **2,** 11 (1956).
7. D. F. Shriver, "The Manipulation of Air-sensitive Compounds," McGraw-Hill Book Company, New York, 1969.
8. M. L. Morris, R. W. Moshier, and R. E. Sievers, *Inorg. Chem.,* **2,** 411 (1963).
9. F. A. Cotton and R. H. Holm, *J. Am. Chem. Soc.,* **82,** 2979 (1960).

23. AMMONIUM PENTAHALOOXOMOLYBDATES(2−)

Submitted by H. K. SAHA* and A. K. BANERJEE*
Checked by SALLY M. HORNER† and GUY O. SIMPSON‡

In recent years oxomolybdenum(V) compounds have been a subject of wide interest to chemists. In many cases, ammonium pentachlorooxo- and pentabromooxomolybdates(V) are used as starting materials for the synthesis of other oxomolybdenum(V) compounds. A review of the literature[1-4] reveals the need for an easy and convenient chemical method for reduction of hexavalent molybdenum to the pentavalent state. Reduction by metallic mercury[2] suffers from serious difficulties in that it is time-consuming and costly and involves troublesome separation of solid mercurous chloride. However, we have modified the hydrogen halide reduction method devised by Allen et al.[3] and studied kinetically by Mauro and Daneshi[5] to the point that it is a rapid and convenient method for synthesis of pentahalooxomolybdate(2−) salts.[6,7]

*Inorganic Chemistry Laboratory, University College of Science, 92 Acharya Prafulla Chandra Road, Calcutta 9, India.
†Department of Chemistry, Meredith College, Raleigh, N.C.
‡Department of Chemistry, Methodist College, Fayetteville, N.C.

A. AMMONIUM PENTACHLOROOXOMOLYBDATE(2−)

$$2(NH_4)_2[MoO_4] + 2HI + 10HCl \longrightarrow 2(NH_4)_2[MoOCl_5] + I_2 + 6H_2O$$

Procedure

■ **Caution.** *This preparation should be carried out in an efficient fume hood!*

About 2 g. of reagent-grade ammonium paramolybdate is treated with 15 ml. of concentrated hydrochloric acid, and the mixture is heated to boiling with constant stirring. Concentrated hydriodic acid (2 ml.) is added, and the solution is carefully boiled to drive off iodine and finally is evaporated to a sticky, moist solid. This solid is redissolved in 15 ml. of concentrated hydrochloric acid, and the solution is evaporated again. Finally, the moist solid is dissolved by boiling in 15 ml. of hydrochloric acid to give a bright green solution. This solution is concentrated by slow boiling to a volume of 7–8 ml. and is cooled in a freezing mixture; then gaseous hydrogen chloride is added to saturate the solution. Bright green crystals appear and are filtered and dried in a vacuum desiccator over solid KOH to constant weight; yield is 2.5 g. *Anal.* Calcd. for $(NH_4)_2[MoOCl_5]$: Mo, 29.47; Cl, 54.54; N, 8.61. Found: Mo, 29.44; Cl, 53.81; N, 8.91.

Analysis. The compound was treated with Na_2O_2 to oxidize molybdenum to the hexavalent state, and then Na_2S was added to convert it to thiomolybdate. Molybdenum(VI) sulfide was precipitated by acidification and finally ignited at 510° to MoO_3. Halogen was estimated in the filtrate and washings as silver halide. Nitrogen was estimated by the Dumas method. Oxidation state was determined by the ceric sulfate method[8] to be 4.97. The magnetic moment, μ_{eff} 1.80 B.M. at 303°K., was determined by the Gouy balance method.

The checkers analyzed for molybdenum by dissolving the sample in 5% H_2SO_4 and reducing to Mo(III) in a Jones reductor (30-mesh zinc amalgam). The reduced solution was received directly in acidic iron(III) sulfate solution and was titrated with a standard cerium(IV) solution with ferroin as the indicator.

Properties

The compound is freely soluble in water, giving a reddish-brown solution, which reacts acidic to litmus, owing to hydrolysis of the ion $[MoOCl_5]^{2-}$. When ammonia or alkali is added to this solution, a brown precipitate of $MoO(OH)_3$ is obtained. With anhydrous alcohol it precipitates ammonium chloride, leaving a brown solution containing solvated $MoOCl_3$.

B. AMMONIUM PENTABROMOOXOMOLYBDATE(2−)

$$2(NH_4)_2[MoO_4] + 12HBr \longrightarrow 2(NH_4)_2[MoOBr_5] + Br_2 + 6H_2O$$

Procedure

■ **Caution.** *This preparation should be carried out in an efficient fume hood!*

A sample of 2 g. of ammonium para-molybdate is treated with 20 ml. of 9 M hydrobromic acid and is gently boiled to drive off the liberated bromine. The solution is evaporated almost to dryness. The process is repeated with a second 20-ml. portion of hydrobromic acid. Finally the residue is dissolved in 10 ml. of 9 M hydrobromic acid. The solution is concentrated to half volume and cooled in a freezing mixture. Anhydrous hydrogen bromide* gas is rapidly passed through this solution to precipitate reddish-brown, crystalline ammonium pentabromooxomolybdate(2−). This solid is filtered and dried in a vacuum desiccator over solid KOH to constant weight; yield is 3 g. *Anal.* Calcd. for $(NH_4)_2[MoOBr_5]$: Mo, 17.52; Br, 73.00; N, 5.11. Found: Mo, 18.01; Br, 72.51; N, 5.48. The oxidation state of the metal was found to be 4.99, and the magnetic moment $\mu_{eff} = 1.79$ B.M. at 303°K.

Properties

The compound dissolves readily in water, and the aqueous solution reacts acidic to litmus owing to hydrolysis. Brown $MoO(OH)_3$ is precipitated from this solution on adding ammonia

---

*Available from the Matheson Company, Inc., East Rutherford, N.J. 07073.

or alkali to it. When the compound is treated with anhydrous alcohol, ammonium bromide is precipitated, leaving a brown solution of solvated $MoOBr_3$.

References

1. R. G. James and W. Wardlaw, *J. Chem. Soc.,* **1927**, 2145.
2. W. G. Palmer, "Experimental Inorganic Chemistry," p. 418, Cambridge University Press, New York, 1954.
3. J. F. Allen and H. M. Neumann, *Inorg. Chem.,* **3**, 1612 (1964).
4. J. P. Simon and P. Souchay, *Bull. Soc. Chim. France,* **1956**, 1402.
5. F. Mauro and L. Daneshi, *Gazz. Chim. Ital.,* **11**, 286 (1881).
6. M. C. Halder and H. K. Saha, *J. Indian Chem. Soc.,* **44**, 231 (1967).
7. H. K. Saha and A. K. Banerjee, *ibid.,* **45**, 660 (1968).
8. W. M. Charmichael and D. A. Edwards, *J. Inorg. Nucl. Chem.,* **1968**, 2641.

24. 13-VANADOMANGANATE(IV) AND NICKELATE(IV)

Submitted by GEORGE B. KAUFFMAN,* RUSSELL FULLER,* JAMES FELSER,* CHARLES M. FLYNN, JR.,† and MICHAEL T. POPE†
Checked by LOUIS C. W. BAKER†

The heteropoly anions of Group VIB elements (chromium, molybdenum, and tungsten) have been extensively investigated, but those of Group VB elements (vanadium, niobium, and tantalum) have been somewhat neglected. Heteropolyniobates of manganese(IV) and nickel(IV),[1-3] a manganese(IV) heteropoly compound with both niobium and tantalum,[3] and chromium(III) and cobalt(III) complexes with ethylenediamine and hexaniobate have been characterized recently.[4] The 13-vanadomanganate(IV) and nickelate(IV) are the first heteropoly anions with the unusual 1:13 stoichiometry. The compounds have been prepared in good yield by the slow reaction of manganese(II) or nickel(II) ion, isopolyvanadate ion, and a moderate excess of peroxydisulfate ion in the pH range 3–5.[5]

*California State College at Fresno, Fresno, Calif. 93710. Financial support of the donors of the Petroleum Research Fund administered by the American Chemical Society (Grant 1152-B), the National Science Foundation (Grants GS-1580 and GY-2607), the Research Corporation, the California State College at Fresno Research Committee, and the U.S. Air Force Office of Scientific Research (Grant AFOSR-1066-66) is gratefully acknowledged.
†Georgetown University, Washington, D.C. 20007.

At a pH near 6, manganese(II) metavanadate precipitates and does not react further, whereas at a pH near 2, other heteropoly species are formed. Inasmuch as commercial potassium metavanadate is not suitable for the syntheses of the 13-vanadomanganate(IV) and nickelate(IV), directions for the preparation of this compound are included in the syntheses below, which are modified from the method of Flynn and Pope.[5] The preparation of the yellow-tinged green reduced form of potassium 13-vanadomanganate(IV) is also included.[5]

A. POTASSIUM VANADATE(V)

(Potassium Metavanadate)

$$K_2CO_3 + V_2O_5 \longrightarrow 2K[VO_3] + CO_2$$

Procedure

Potassium carbonate (69.11 g., 0.50 mole) is dissolved with mechanical stirring in 400 ml. of water, contained in an 800-ml. beaker, and maintained at 85–90° by means of a water bath or well-adjusted hot plate. To the clear, colorless solution is added with heating and constant stirring 90.95 g. (0.500 mole) of vanadium-(V) oxide in small portions (■ **Caution.** *Effervescence*), which results in a tan to dark-brown opaque mixture. When the effervescence subsides, 10 ml. of 3% hydrogen peroxide is added to assist dissolution of the vanadium oxide by oxidizing the small proportion of vanadium(IV) that is present in the commercial product. The reaction mixture is maintained at 85–90° with constant stirring for about 2 1/2 hours, at which time the volume has been reduced to 150–200 ml. by evaporation. The mixture is heated to just below boiling and is filtered immediately by suction through a 9-cm. coarse-porosity, fritted-glass funnel (the mixture is too viscous to be filtered through filter paper) into a 500-ml. filter flask. The coarse brown to black residue is discarded. A boiling chip is added to the filter flask containing the clear, light-yellow, slightly viscous filtrate. The exact color of the solution depends upon the pH; if it is deep yellow, too little K_2CO_3 has been used; if it is colorless, too much K_2CO_3 has been used. The flask is stoppered and evaporated with aspirator suction on an 85–90°C. water bath. (Use of a hot plate may decompose the product.) The solution becomes increasingly viscous, and suddenly a nearly white precipitate begins to form. When the volume of the mixture has been reduced to about 75 ml., the powdery, pale-yellow precipitate is collected on a 9-cm. fritted-glass funnel (*coarse-porosity*), washed first with 20 ml. of ice water, then with 20 ml. of 95% ethanol, and finally air-dried. The yield of powdery, pale-yellow product is about 40 g. (*ca.* 29%). If the reaction mixture is evaporated to a volume of about 25 ml., a second crop of product (about 30 g., *ca.* 22%) may be obtained. *Anal.*[5] Calcd. for KVO_3: V, 36.9. Found: V, 36.8, 37.1.

Properties

Commercial potassium metavanadate may contain significant amounts of more basic vanadates, $K_4[V_2O_7]$ and $K_3[VO_4]$. It gives low yields when used as starting material for the syntheses of $K_7[MnV_{13}O_{38}] \cdot 18H_2O$ and $K_7[NiV_{13}O_{38}] \cdot 16H_2O$. The commercial product dissolves readily in water to yield a colorless solution of pH about 10. Pure potassium metavanadate is colorless, but the product prepared here is usually pale yellow because of traces of intense yellow-orange $K_6[V_{10}O_{23}]$ formed by reaction with atmospheric carbon dioxide. It dissolves in water with difficulty to yield a pale-yellow solution of pH about 7. The species of vanadium(V) present in solution is a function of pH.[6]

B. POTASSIUM 13-VANADOMANGANATE(IV) 18-HYDRATE

$$13K[VO_3] + MnSO_4 \cdot H_2O + K_2S_2O_8 + H_2SO_4 + 16H_2O \longrightarrow$$

$$K_7[MnV_{13}O_{38}] \cdot 18H_2O + 4K_2SO_4$$

Procedure

In 500 ml. of water, maintained at 80° in an 800-ml. beaker by means of a water bath, 17.95 g. (0.130 mole) of potassium metavanadate (prepared according to Sec. A) is dissolved by continuous mechanical stirring. To the resulting light-yellow solution (pH *ca.* 7 ± 0.5) 10 ml. of 0.5 *M* sulfuric acid is added, whereupon the color becomes red or deep orange. One-hundredth mole of manganese(II) sulfate (1.69 g. of the monohydrate or a corresponding weight of another hydrate)* and 5.40 g. (0.020 mole) of potassium peroxy-disulfate are next added in that order, with no resulting visible change. The stirring and heating are continued. The solution slowly becomes reddish-brown in color, and after about an hour some orange precipitate has formed. When the mixture has evaporated to a volume of about 150 ml. (5–7 hours), the opaque brown-black mixture is heated on a hot plate almost to boiling, with vigorous stirring, and is *immediately* filtered with suction through paper on a 9-cm. Büchner funnel. (A sintered-glass funnel will clog and allow passage of the precipitate into the filtrate.) The small amount of residue, which consists of orange-brown and dark-brown layers, is discarded.

To the opaque, almost black, filtrate is added 20 ml. of 1 M potassium acetate, and the filtrate is reheated almost to boiling in order to redissolve the precipitate that has begun to form as the solution cools. The solution is allowed to cool slowly to room temperature. (An ice bath should not be used, because it will cause contamination of the product by impurities which are soluble at room temperature.) The resulting lustrous, red-orange crystals are collected by suction on a 9-cm. Büchner funnel, rinsed with 20 ml. of 0.5 M potassium acetate–0.5 M acetic acid, and air-dried. The yield of crude product is *ca.* 14.5 g. (*ca.* 75%). The product is purified by dissolving in a minimum volume (*ca.* 200 ml.) of 0.5 M potassium acetate–0.5 M acetic acid which has been heated just below boiling, allowing the solution to cool slowly to room temperature (not lower), and collecting the recrystallized product by suction on a 9-cm. Büchner funnel. The bright red-orange octahedral crystals are washed with 20 ml. of 1 : 1 ethanol-water, 20 ml. of 95% ethanol, and air-dried. The yield is *ca.* 13.5 g. (*ca.* 70%). *Anal.*[5] Calcd. for $K_7[MnV_{13}O_{38}] \cdot 18H_2O$: K, 14.2; Mn, 2.86; V, 34.4; H_2O, 16.9. Found: K, 14.2; Mn, 2.88; V, 34.8; H_2O, 16.9.

C. REDUCED POTASSIUM 13-VANADOMANGANATE(IV)

$$K_7[Mn^{IV}V_{13}{}^VO_{38}] \cdot 18H_2O + K_4[Fe(CN)_6] + 1/2H^+ \longrightarrow$$

$$K_{7.5}H_{0.5}[Mn^{IV}V_{12}{}^VV^{IV}O_{38}] \cdot 17H_2O + K_3[Fe(CN)_6] + 1/2K^+ + H_2O$$

---

*If commercial potassium metavanadate has been used, precipitation occurs at this point, and only a poor yield of product will be obtained.

Procedure

To 90 ml. of water contained in a 250-ml. beaker is added 2.89 g. (0.0015 mole) of potassium 13-vanadomanganate(IV) 18-hydrate (Sec. B), 30 ml. of 1 *M* potassium acetate, and 30 ml. of 1 *M* acetic acid. To the mechanically stirred solution is added 15 ml. of 0.1 *M* potassium hexacyanoferrate(II) (potassium ferrocyanide), whereupon the orange solution becomes dark iridescent, gold-green, and a lustrous, yellow-tinged green precipitate begins to form. After the mixture has been stirred for about 5 minutes, the olive-green product is collected by suction filtration on a 5-cm. Büchner funnel, washed successively with 10 ml. of 0.2 *M* potassium acetate–0.2 *M* acetic acid, 10 ml. of 1:1 ethanol-water, and 10 ml. of 95% ethanol, and then air-dried. The yield is 2.14 g. (74%). *Anal.*[5] Calcd. for $K_{7.5}H_{0.5}$-$[MnV_{13}O_{38}] \cdot 17H_2O$: K, 15.2; Mn, 2.85; V, 34.4; H_2O, 16.1. Found: K, 15.0; Mn, 2.82; V, 34.5; H_2O, 15.9.

D. AMMONIUM 13-VANADOMANGANATE(IV) 18-HYDRATE AND 5-HYDRATE

$$13[NH_4][VO_3] + MnSO_4 \cdot H_2O + (NH_4)_2[S_2O_8] + H_2SO_4 +$$

$$16H_2O \longrightarrow [NH_4]_7[MnV_{13}O_{38}] \cdot 18H_2O + 4[NH_4]_2SO_4$$

Procedure

The procedure is the same as that for potassium 13-vanado-manganate(IV) (Sec. B), with the substitution of commercial ammonium metavanadate (15.20 g., 0.130 mole) for potassium metavanadate and ammonium peroxydisulfate (4.56 g., 0.020 mole) for potassium peroxydisulfate. At the end of the reaction (*ca.* 6 hours, volume of reaction mixture *ca.* 75 ml.), after the reaction mixture has been filtered,* 1.54 g. (0.020 mole) of ammonium acetate is added with stirring. The filtrate is transferred to a 9-cm. evaporating dish and set aside for *ca.* 3 days or until the volume has decreased to *ca.* 25 ml. The crystals of bright red crude product are collected by suction filtration on a 9-cm. Büchner funnel and air-dried (*ca.*

---

*Heating the reaction mixture almost to boiling before removal of waste material is unnecessary.

10.5 g., *ca.* 68%). The crude product is purified by recrystallizing from a minimum volume (*ca.* 75 ml.) of hot 0.5 M ammonium acetate–0.5 M acetic acid. The solution is allowed to stand in a 9-cm. evaporating dish for *ca.* 3 days or until the volume has decreased to *ca.* 25 ml. The bright red-orange crystals of product are collected by suction filtration on a 9-cm. Büchner funnel, washed with 20 ml. of ice cold 1:1 ethanol-water, blotted dry with tissue paper, and stored in the refrigerator in a tightly closed container. The 18-hydrate is converted to the 5-hydrate by repeated washing with 95% ethanol at room temperature, during which process the crystals turn opaque and crumble to a powder. The yield is *ca.* 61% (*ca.* 9.4 g. for the 18-hydrate or *ca.* 8.2 g. for the 5-hydrate). *Anal.*[5] Calcd. for $[NH_4]_7[MnV_{13}O_{38}] \cdot 5H_2O$: NH_4, 8.19; Mn, 3.56; V, 42.9; H_2O, 5.8. Found: NH_4, 8.29; Mn, 3.60; V, 42.7; H_2O, 6.1.

E. POTASSIUM 13-VANADONICKELATE(IV) 16-HYDRATE

$$13K[VO_3] + NiSO_4 \cdot 6H_2O + K_2[S_2O_8] + H_2SO_4 + 9H_2O \longrightarrow$$
$$K_7[NiV_{13}O_{38}] \cdot 16H_2O + 4K_2SO_4$$

Procedure

The procedure is the same as that for potassium 13-vanado-manganate(IV) (Sec. B), with the substitution of nickel(II) sulfate (2.63 g., 0.01 mole, of the hexahydrate or a corresponding weight of another hydrate) for manganese(II) sulfate and the substitution of acetone for ethanol. Inasmuch as potassium 13-vanadonickelate-(IV) is less soluble than the corresponding 13-vanadomanganate(IV), the Büchner funnel should be preheated by pouring boiling water through it before the reaction mixture is filtered (Sec. B, Paragraph 1). The yield of crude product is *ca.* 12.1 g. (64%), and that of the recrystallized black, octahedral crystals is *ca.* 11.0 g. (58%). *Anal.*[5] Calcd. for $K_7[NiV_{13}O_{38}] \cdot 16H_2O$: K, 14.5; Ni, 3.10; V, 35.0; H_2O, 15.2. Found: K, 14.6; Ni, 3.10; V, 34.8; H_2O, 14.8.

Properties

The 13-vanadomanganate(IV) and nickelate(IV) ions have been characterized by complete analyses of several salts, cryoscopic molecular-weight determinations in saturated sodium sulfate solutions,[7] magnetic-susceptibility measurements, and preliminary x-ray crystallographic examination. The salts $K_7[MnV_{13}O_{38}] \cdot 18H_2O$ and $K_7[NiV_{13}O_{38}] \cdot 16H_2O$ are isomorphous. The effective magnetic moment of the manganese compound is 3.91 B.M. in close agreement with the value 3.88 B.M. calculated for a high-spin d^3 (Mn(IV)) configuration. The nickel compound is diamagnetic.

The solid salts are stable indefinitely, although the higher hydrates of the sodium and ammonium salts are very efflorescent, and the potassium salts are slightly efflorescent. In solution, both the manganate and nickelate are most stable in the approximate pH range 3–6. At a pH above 6, solutions of both complexes darken in color and ultimately decompose to isopolyvanadates with precipitation of the manganese or nickel as dark-brown or black solids, respectively. The solubilities of the manganate and nickelate salts of potassium are in the range $0.03-0.04M$ in water and $0.005-0.01M$ in $0.5M$ potassium acetate–$0.5M$ acetic acid. The manganate complex is precipitated by cesium, barium, guanidinium and silver ions, but not by mono-, di-, tri-, or tetramethylammonium ions. Dehydration studies suggest the presence of constitutional water in the manganate complexes. A proposed structure[5] for the anion in $K_7[MnV_{13}O_{38}] \cdot 18H_2O$, consistent with the observed unit-cell symmetry, requires 39 oxygen atoms (one constitutional water molecule).

The manganate ion is not reduced by bromide ion but is reduced slowly by iodide ion and quickly by vanadyl(IV) or hexacyanoferrate(II) ions. When the latter two ions are used as reductants, especially with the potassium complex, green products are obtained rapidly and in high yield. The green species is unstable in solution and is apparently in equilibrium with the reactants. With potassium salts, the solubility of the product is low, and the reaction is driven to completion. Potentiometric titrations show that a one-electron reduction occurs to produce the green species, which has been characterized by analysis and optical and e.s.r. spectroscopy. It is a mixed-valence species similar to the heteropoly blues of molybdenum and tungsten. E.s.r. spectra suggest that the extra electron is fairly well trapped on a specific vanadium atom, and the complex is therefore a class II mixed-valence species.[8]

The nickelate is reduced by iodide ion and by hydrogen peroxide, but the reactions are not immediate at room temperature. Vanadyl-

(IV) and hexacyanoferrate(II) ions reduce the nickel(IV) complex quickly to nickel(II) but with no evidence for formation of species analogous to the green reduced manganate.

References

1. B. W. Dale and M. T. Pope, *Chem. Commun.,* **1967,** 792.
2. C. M. Flynn, Jr., and G. D. Stucky, *Inorg. Chem.,* **8,** 332, 335 (1969).
3. B. W. Dale, J. M. Buckley, and M. T. Pope, *J. Chem. Soc. (A),* **1969,** 301.
4. C. M. Flynn, Jr., and G. D. Stucky, *Inorg. Chem.,* **8,** 178 (1969).
5. C. M. Flynn, Jr., and M. T. Pope, *J. Am. Chem. Soc.,* **92,** 85 (1970).
6. M. T. Pope and B. W. Dale, *Quart. Rev. (London),* **22,** 527 (1968).
7. G. A. Tsigdinos, Ph.D. dissertation, pp. 81–88, Boston University, 1961.
8. M. B. Robin and P. Day, *Advan. Inorg. Chem. Radiochem.,* **10,** 247 (1967).

Chapter Five
BORON COMPOUNDS

25. OCTAHYDROTRIBORATE (1 −) ([B$_3$H$_8$]$^-$) SALTS

Octahydrotriborate $(1 -)$ ion, $[B_3H_8]^-$, is an important intermediate in the synthesis of higher boranes, polyhedral borane anions, and transition-metal complexes. Salts containing this ion have usually been prepared[1] from diborane, which is toxic, spontaneously flammable, and expensive. In addition, pressure facilities are often required for the synthesis.

Two new syntheses avoid these complications by generation of diborane *in situ.* As described below, the reactions of sodium tetrahydroborate with iodine[2] (Method A) or boron trifluoride[3] (Method B) in diglyme solution afford solutions of $Na[B_3H_8]$ in good yield with relatively little effort. The sodium salt may be precipitated as a dioxane solvate as illustrated in Method B or may be directly converted to a tetraalkylammonium salt as in Method A. The hydrophobic alkyl groups in the tetrabutylammonium salt provide good organic solubility. The solvated sodium salt is very water-soluble, but its stoichiometry is variable and makes it unsuitable for some reactions. In such situations it may be easily converted to the moderately soluble cesium or tetramethylammonium salts as described in Method B.

Purification of Solvent

In both syntheses, a vital step is purification of the solvent, bis-(2-methoxyethyl) ether, generally referred to as diglyme (or, occasionally, as diethylene glycol dimethyl ether). This solvent, as obtained from Ansul Chemical Company, seems to contain a high-boiling alcohol, perhaps the monomethyl ether, as an impurity. In spite of a very effective oxidation inhibitor, small amounts of peroxide are often present. Both types of impurity are removed by a pretreatment with calcium hydride or sodium, but distillation from an aluminum hydride complex salt is required to make the solvent sufficiently anhydrous for the $Na[B_3H_8]$ synthesis. The procedure described here utilizes the hazardous[4] but commonly available lithium tetrahydridoaluminate as the final drying agent. Vitride reagent $(Na[AlH_2(OCH_2CH_2OCH_3)_2])$, now available from Eastman Organic Chemicals, appears to serve the same function and does not spontaneously ignite in moist air. When $Li[AlH_4]$ is used as the final drying agent, the distillation[5] must be conducted under a vacuum such that the distillation-flask temperature is well below the decomposition temperature (125°) of $Li[AlH_4]$.

For preliminary purification, the diglyme is stored over CaH_2 for 24 hours and then decanted from the drying agent. Alternatively, as suggested by Hawthorne and Leyden, the diglyme is refluxed over molten sodium for 4 hours and is then distilled from the sodium at atmospheric pressure (b.p. 161°).

Lithium tetrahydridoaluminate is then added *slowly and cautiously* to the predried diglyme until gas evolution subsides. (■ **Caution.** *The reaction is very exothermic.*) Some excess hydride is left suspended in the diglyme. After 12 hours the mixture is decanted or filtered through glass wool into a distillation flask, and a *small* amount of $Li[AlH_4]$ is added for distillation (avoid excess). The distillation should be done in a hood behind a suitable safety shield, and it should be stopped before the distilling mixture becomes dry. At 15 mm. pressure, diglyme distills at 62–63°. It should be stored under nitrogen until use. The slurry of $Li[AlH_4]$ in diglyme remaining in the distillation flask should be *cautiously* treated with moist diglyme until all reaction ceases before addition of gross quantities of water.

METHOD A

$$3Na[BH_4] + I_2 \longrightarrow 2NaI + 2H_2 + Na[B_3H_8]$$
$$Na[B_3H_8] + [(n\text{-}C_4H_9)_4]I \longrightarrow [(n\text{-}C_4H_9)_4N][B_3H_8] + NaI$$

Submitted by G. E. RYSCHKEWITSCH* and K. C. NAINAN*
Checked by S. R. MILLER† and L. J. TODD†

■ **Caution.** *The reaction should be carried out in a well-ventilated hood behind a safety shield and away from flames or spark sources, since the evolved hydrogen is an explosion or fire hazard. The boranes that might escape from the bubbler or which could be produced during the precipitation of octahydrotriborate are toxic and spontaneously flammable.*

Procedure

In a dry-box (or a nitrogen atmosphere) 17 g. (0.45 mole) of powdered sodium tetrahydroborate is slurried with 250 ml. of anhydrous diglyme in a 1-l. three-necked flask. The flask is equipped with a sealed mechanical stirrer in the central neck, and one of the other two necks is fitted with a 125-ml. pressure-compensating dropping funnel whose tip is extended below the surface of the mixture. The third neck is connected to a bubbler containing a benzene-amine (4-picoline was used) mixture to scrub the gaseous boranes evolving from the reaction as minor products. During assembly a gentle stream of dry nitrogen is maintained through the apparatus to minimize the possible entrance of air into the system. After rapidly transferring a solution of iodine into the funnel (20.6 g., 81.0 mmoles in 115 ml. of dry diglyme), the entire reaction system is purged with nitrogen for about 10 minutes, and the reaction flask is placed in a previously heated oil bath whose temperature is adjusted to 98–102° before the addition of iodine. After the nitrogen flow is stopped, the iodine solution is added dropwise during a

*University of Florida, Gainesville, Fla. 32601.
†Indiana University, Bloomington, Ind. 47401.

period of $1\frac{1}{4}$–$1\frac{1}{2}$ hours, to the hot, vigorously stirring reaction mixture. The hydrogen gas produced passes through the bubbler. The reaction mixture is stirred for two more hours while the temperature is maintained at about 95°. The volume is then reduced to 145 ml. by passing dry nitrogen over the warmed reaction mixture at 50°, or alternately by pumping. The cooled mixture, together with washings of 50 ml. of water, is transferred to a 2-l. beaker. About 1 l. of saturated aqueous tetra-*n*-butylammonium iodide is added slowly with vigorous stirring until no more precipitation takes place.* The white precipitate is filtered on a Büchner funnel, washed with water (about 900 ml.), and dried under vacuum. The yield is 13.42 g., or 59%, based on iodine used. M.p. 208.5–210.5°, with decomposition. If octahydrotriborate(1−) is precipitated after complete removal of diglyme, a higher yield of the crude product with somewhat lower melting point may be obtained.

A portion of the crude salt (5.21 g.) is dissolved in 30 ml. of methylene chloride (dichloromethane), filtered, and washed with 30 ml. of methylene chloride, and reprecipitated by adding 400 ml. of diethyl ether. The precipitate after drying *in vacuo* weighs 4.45 g. (86% recovery). M.p. 210–212°, with decomposition.

The tetrabutylammonium salt can be converted readily to the bis(triphenylphosphino)methylium salt, $[[(C_6H_5)_3P]_2CH]^+[B_3H_8]^-$. A solution of 0.231 g. of $[(n\text{-}C_4H_9)_4N][B_3H_8]$ in 10 ml. of absolute ethanol is mixed with 0.5 g. of $[[(C_6H_5)_3P]_2CH]$ Br[6] in 6 ml. of the same solvent. The precipitate, washed with 10 ml. of alcohol and vacuum-dried, weighs 0.385 g. (82% yield). An analytical sample is obtained as above by precipitation with ether from methylene chloride solution. M.p. 243–244°, with decomposition. In a similar fashion the μ-nitrido-bis[triphenylphosphorus](1+) salt, $[(C_6H_5)_3P]_2N]^+[B_3H_8]^-$, is obtained. From 1.00 g. $[(n\text{-}C_4H_9)_4N]$-$[B_3H_8]$ in 50 ml. of ethanol and 2.14 g. $[[(C_6H_5)_3P]_2N]Cl$[7] in 10 ml. of ethanol, 1.43 g. (70% yield) of dry iminium salt was precipitated. A reprecipitated sample softened at 221° and melted at 230–232°, with decomposition.

---

*The reaction mixture obtained after partial evaporation of diglyme reacts vigorously with water, and the odor of possibly toxic boranes can be noted. Therefore, the addition of water or aqueous solution of tetra-*n*-butylammonium iodide to the reaction mixture must be done slowly and in a hood.

METHOD B

$$5Na[BH_4] + 4BF_3 \cdot O(C_2H_5)_2 \longrightarrow$$
$$3Na[BF_4] + 2Na[B_3H_8] + 2H_2 + 4(C_2H_5)_2O$$
$$Na[B_3H_8] + 3C_4H_8O_2 \longrightarrow Na[B_3H_8] \cdot 3(C_4H_8O_2)$$

Submitted by W. J. DEWKETT,* M. GRACE,* and H. BEALL*
Checked by M. F. HAWTHORNE† and R. LEYDEN†

■ **Caution.** *The reaction should be carried out in a well-ventilated hood behind a safety shield and away from flames or spark sources, since the evolved hydrogen is an explosion or fire hazard. The boranes that might escape from the bubbler or which could be produced during the precipitation of octahydrotriborate are toxic and spontaneously flammable.*

Procedure

Ten grams (0.26 mole) of sodium tetrahydroborate is dissolved in 500 ml. of diglyme. The large quantity of diglyme is used because of the very limited solubility of $Na[BH_4]$ in the solvent at $100°$.[8] The solution is placed in a 1-l. three-necked flask which is equipped with a nitrogen inlet in one neck, a nitrogen outlet in the second, and a pressure-equalizing funnel, the tip of which is extended below the surface of the $Na[BH_4]$ solution, in the third. Then a slow stream of N_2 is allowed to pass through the system while the solution is heated to $40°$ for maximum dissolution of the $Na[BH_4]$. The effluent gas from the system is bubbled through about 100 ml. of acetone or of a solution of pyridine in benzene in order to destroy any diborane which may be formed in the reaction. Care must be taken to avoid return of the solution or of air to the reaction vessel.

*Department of Chemistry, Worcester Polytechnic Institute, Worcester, Mass. 01609. This work was supported by the Petroleum Research Fund administered by the American Chemical Society. W. J. D. acknowledges the National Science Foundation for a predoctoral traineeship.

†Department of Chemistry, University of California at Los Angeles, Los Angeles, Calif. 90024.

Under a stream of nitrogen 16.0 ml. (0.13 mole) of pure boron trifluoride etherate* is then mixed with 100 ml. of diglyme and placed in the dropping funnel. Before addition of the $BF_3 \cdot O(C_2H_5)_2$, the nitrogen flow is stopped and the inlet port is sealed from the nitrogen supply by a pinch clamp. With vigorous stirring of the $Na[BH_4]$ solution by a magnetic stirrer, the $BF_3 \cdot O(C_2H_5)_2$ is added dropwise over a period of 30–60 minutes. The solution is then heated to 100° and maintained at this temperature for 2 hours. The reaction mixture is then allowed to cool to room temperature, during which time nitrogen is again allowed to flow through the system in order to remove any residual B_2H_6 and to ensure against scrubbing-solution backup during cooling.

The cooled solution is filtered in air to remove $Na[BF_4]$, and the diglyme is removed by distillation at 60° under vacuum. This diglyme is ready for drying as detailed above and then for reuse in subsequent preparations. The $Na[B_3H_8]$ is separated from the crude product by repeated extraction with 150-ml. portions of anhydrous ethyl ether and is isolated as the dioxane adduct by addition of enough dioxane to each ether portion to effect precipitation of the insoluble $Na[B_3H_8] \cdot 3(C_4H_8O_2)$. Ether extraction may be terminated when this precipitate is no longer formed. The solid product is removed by filtration and dried in vacuum to yield 13.4 g. of $Na[B_3H_8] \cdot 3(C_4H_8O_2)$ [63% yield based on $BF_3 \cdot O(C_2H_5)_2$]. (■ **Caution.** *Do not draw air through the solid or solution, because such materials have occasionally inflamed spontaneously.*)

The identity of the product may be confirmed by its infrared and n.m.r. spectra, since the stoichiometry of the dioxanate is variable. The infrared spectrum shows prominent bands at *ca.* 2440, 2400, 2370, 2320, 2130, 2080, 1140, and 1010 cm.$^{-1}$. The ^1H n.m.r. of the product in D_2O gave the expected decet for the $[B_3H_8]^-$ ion with $J_{B-H} = 33$ Hz. Conversion to the cesium salt gives a moderately stable material of constant composition suitable for analysis:

$$CsBr + Na[B_3H_8] \cdot 3(C_4H_8O_2) \longrightarrow$$

$$NaBr + Cs[B_3H_8] + 3C_4H_8O_2$$

---

*The colored commercial material (Eastman or Matheson, Coleman, and Bell) should be distilled *in vacuo* to obtain colorless $BF_3 \cdot O(C_2H_5)_2$.

Cesium bromide (2.0 g., 9.4 mmoles) is dissolved in 10 ml. of water, and to this is added 3.0 g. of $Na[B_3H_8] \cdot 3(C_4H_8O_2)$ (9.2 mmoles). (■ **Note.** *The order of addition is important!*) An additional 10 ml. of water is added, and the solution is cooled to ice-bath temperature. The $Cs[B_3H_8]$ precipitates from solution and is filtered and dried under vacuum. One gram of $Cs[B_3H_8]$ is recovered, corresponding to a 63% yield based on $Na[B_3H_8] \cdot 3(C_4H_8O_2)$. *Anal.* Calcd. for CsB_3H_8: Cs, 76.6; B, 18.7; H, 4.6. Found: Cs, 76.5; B, 18.6; H, 4.3.

Properties

Salts of the $[B_3H_8]^-$ ion are white, nonvolatile solids possessing varying degrees of stability dependent upon the nature of the cation and the degree of solvation. The nonsolvated salts such as $Cs[B_3H_8]$ and $[R_4N][B_3H_8]$ are the most stable. The structure of the $[B_3H_8]^-$ ion was determined by an x-ray crystal study of $[H_2B(NH_3)_2][B_3H_8]$.[9]

$[(n-C_4H_9)_4N][B_3H_8]$ is a white crystalline solid which is readily soluble in methylene chloride or acetonitrile, but only sparingly soluble in benzene, diethyl ether, saturated hydrocarbons such as hexane, or water. The infrared spectrum of tetra-*n*-butylammonium octahydrotriborate in a potassium bromide pellet shows the following prominent bands associated with B—H vibrations: 2450/2400 (doublet, s), 2310 (sh), 2120/2080 (doublet, m), 1140 (s), and 1015 (s) cm.$^{-1}$. The ^{11}B n.m.r. spectrum of the compound in CH_2Cl_2 is a septet, $J_{B-H} = 33$ Hz., centered at $+47.9$ p.p.m. relative to external trimethyl borate. The 60-MHz. ^1H n.m.r. spectrum of the μ-nitrido-bis[triphenylphosphorus] (1 +) salt in methylene chloride–d_2 shows the expected 10 peaks, centered at $\tau 9.73$, $J_{B-H} = 32$ Hz. In the tetra-*n*-butylammonium salt the down-field half of the decet is masked by the absorption of the cation.

References

1. (*a*) W. Hough, L. Edwards, and A. McElroy, *J. Am. Chem. Soc.,* **80,** 1828 (1958); (*b*) M. Ford, W. Hough, and L. Edwards, *U.S.A.E.C. Nucl. Sci. Abstr.,* **11,** No. 6233 (1957); (*c*) R. Parry and L. Edwards, *J. Am. Chem. Soc.,* **81,** 3554 (1959); (*d*) G. Kodama and R. Parry, *ibid.,* **82,** 6250 (1960); (*e*) B. Graybill, J. Ruff, and M. F. Hawthorne, *ibid.,* **83,** 2669 (1961); (*f*) D. Gaines, R. Schaeffer,

and F. Tebbe, *Inorg. Chem.*, **2**, 526 (1963); (g) J. Plesek and S. Hermanek, *Coll. Czech. Chem. Commun.*, **31**, 117 (1966); (h) H. C. Miller and E. L. Muetterties, *Inorganic Syntheses*, **10**, 81 (1967).

2. K. C. Nainan and G. E. Ryschkewitsch, *Inorg. Nucl. Chem. Letters*, **6**, 765 (1970).
3. W. J. Dewkett, M. Grace, and H. Beall, *J. Inorg. Nucl. Chem.*, **33**, 1279 (1971).
4. H. R. Watson, *Chem. Ind.*, **1964**, 665.
5. L. F. Fieser and M. Fieser, "Reagents for Organic Synthesis," pp. 255, 583, John Wiley & Sons, Inc., New York, 1967.
6. (a) F. Ramirez, N. B. Desai, B. Hansen, and N. McKelvie, *J. Am. Chem. Soc.*, **83**, 3539 (1961); (b) J. S. Driscoll, D. W. Grisley, Jr., J. E. Pustinger, J. E. Harris, and C. N. Mathews, *J. Org. Chem.*, **29**, 2427 (1964).
7. J. K. Ruff and J. Schlientz, *Inorganic Syntheses*, **15**, 84 (1973).
8. H. C. Brown, E. J. Mead, and B. C. Subba Rao, *J. Am. Chem. Soc.*, **77**, 6209 (1955).
9. C. E. Nordman and C. R. Peters, *ibid.*, **82**, 5758 (1960).

26. PENTABORANE(9) (B_5H_9)

$$[(n\text{-}C_4H_9)_4N][B_3H_8] + HBr \longrightarrow [(n\text{-}C_4H_9)_4N][B_3H_7Br] + H_2$$
$$5[(n\text{-}C_4H_9)_4N][B_3H_7Br] \xrightarrow{100^\circ} 3B_5H_9 + 4H_2 + 5[(n\text{-}C_4H_9)_4N]Br$$

Submitted by V. R. MILLER* and G. E. RYSCHKEWITSCH*
Checked by D. F. GAINES† and N. KEIPE†

Pentaborane(9), since its original synthesis by Stock,[1] has been prepared either by the pyrolysis of diborane or other boron hydrides,[2] or by degradation of B_6H_{12}.[3] The first method suffers from low yields and the inconvenience caused by the presence of B_5H_{11}, which has nearly the same volatility as does B_5H_9, and which must usually be converted to B_5H_9 by lengthy heating. The second method is handicapped by the lack of convenient syntheses for the starting material. The present synthesis[4] proceeds from now readily available starting materials[5] and gives pure product in reasonable yield with little effort. It should be noted that hydrogen halides other than HBr do not give satisfactory results, nor do higher reaction temperatures. Although the first equation corresponds

*Department of Chemistry, University of Florida, Gainesville, Fla. 32601.
†Department of Chemistry, University of Wisconsin, Madison, Wis. 53706.

exactly to the observed reaction stoichiometry, the second equation presents an idealization, because B_2H_6, traces of B_4H_{10}, and a boron-containing solid are also produced.

■ **Editorial Note.** *Pentaborane(9) is an exceptionally reactive compound, and its synthesis presents several hazards, as noted in the Procedure and Properties sections.* The preparation presented here seems to be the best available, but it should be undertaken only by chemists experienced in vacuum-line technique. This preparation seems well suited to the synthesis of isotopically labeled pentaborane (9), for example, $^{10}B_5H_9$. Unlabeled pentaborane(9) is presently commercially available from Callery Chemical Company, Callery, Pa.

Procedure

■ **Caution.** *Diborane and pentaborane are toxic and spontaneously inflame or explode in air. The reaction should be carried out with strict exclusion of air in a shielded apparatus in an efficient fume hood.*

A sample of 9.32 g. (32.9 mmoles) of $(n\text{-}C_4H_9)_4NB_3H_8$ (p. 113), along with a magnetic stirring bar, is placed into a 22-mm.-o.d. Pyrex tube 30 cm. long, which is sealed at one end and attached at the other end to a 1-l. bulb by means of a ground-glass joint. The bulb, in turn, is fitted with a stopcock for evacuation after attachment to a vacuum line capable of operating at less than 10^{-3} torr. Suitable apparatus has been described.[6,7] Dichloromethane (20 ml.), which has been dried by storage over a Type 3A molecular sieve,* is condensed on the solid. After warming to room temperature to dissolve the salt, 32.9 mmoles of hydrogen bromide (purified by passage through a $-78°$ trap) is condensed in the mixture. The mixture is stirred at $-78°$ for 4 hours. Hydrogen gas and the bulk of the solvent are removed by pumping at $-78°$. The product is then thoroughly dried by evacuation at room temperature overnight. (Some diborane is evolved during this period.) Failure to remove all solvent will complicate purification of product and could present an

---

*The molecular sieves should be dried at 100° for several hours and not reexposed to air (preferably not even nitrogen) before use in drying the solvent.

explosion hazard if allowed to contact the pentaborane formed in the next step (see Properties). The solid is then heated to 90–100° for 5 hours with continuous pumping through − 78, − 126, and − 196° traps while the pressure in the bulk is about 1 or 2 torr. Diborane collects in the − 196° trap, a trace of tetraborane (10) and some pentaborane in the − 126° trap, and pentaborane predominantly in the − 78° trap. The B_5H_9 is purified by repeated (*ca.* three times) distillation through a U trap at − 63° (chloroform slush bath) and condensation in a − 95° trap (toluene slush bath). The product should have a vapor pressure of 65 torr at 0°. Pure pentaborane(9) (6.75 mmoles) is obtained in 34% yield according to the idealized equation together with a trace of B_4H_{10} and 7.63 mmoles of diborane. [The checkers obtained 5.64 mmoles (28%) of B_5H_9 and 10.58 mmoles of B_2H_6, possibly contaminated with a little B_4H_{10} and HBr.]

The diborane and the contents of the vacuum-line trap may be diluted with nitrogen and vented into a solution of pyridine in benzene for disposal. (Alternatively pyridine may be condensed in the vacuum-line trap.) Pentaborane(9) is very soluble in the greases normally used as stopcock lubricants. Care must be taken to remove it by evacuation from exposed joints before they are exposed to air.

■ **Caution.** *When large quantities of B_5H_9 are stored at low temperature in a glass container, a solid-state phase change involving expansion can rupture the container.*

Properties

The following data on the properties of pentaborane(9) are extracted from "Pentaborane," Olin Mathieson Chemical Corporation Energy Division Technical Bulletin LF202, 1960. It is strongly recommended that this bulletin be obtained and used as a guide for manipulation of the compound. Pentaborane(9) is toxic and reacts, often with violent explosion, with oxygen. It is immiscible with water and is hydrolyzed only slowly with formation of hydrogen and boric acid. Rapid solvolysis with alcohols such as methanol takes place with hydrogen evolution. Solvents with reactive carbonyl groups, and highly halogenated or oxygenated solvents may form shock-sensitive solutions with pentaborane(9). [■ **Caution.** *Carbon* *tetrachloride, acetone and other ketones, and aldehydes definitely must not be used as solvents for pentaborane(9).*] A list of compounds which the Matheson Company considered potentially dangerous in the presence of pentaborane is:

Carbon tetrachloride	Glycol polyethers such as
Chloroform	diglyme
Methylene chloride	Aldol
Thiokol rubbers	Allyl chloride
Methyl vinyl ketone	Bis(chloromethyl) ether
Ammonium chlorate	3-Chloropropene
Acetylacetone	1,4-Dibromoethane
(2,4-pentanedione)	1,2-Dichloroethane
Crotonaldehyde	Methylene bromide
Dioxane	Methylene iodide
Ethyl acetate	Trichloroethane
Dimethyl ether	Boron trichloride
Acrolein	Freons
Acetone	Trichloroethane

It must be emphasized that any apparatus that has been degreased with trichloroethylene, or any other of the above-mentioned compounds, should be carefully dried to remove the solvent before exposure to pentaborane(9).

Pure pentaborane(9) in dry glass ampuls may be stored for long periods at room temperature. The presence of impurities will result in the generation of pressure by formation of lower boranes and hydrogen.

Pentaborane(9) boils at 58° and freezes at 47°. The vapor pressure of the liquid is represented by the equation

$$\log P_{mm} = 9.96491 - \frac{1951.14}{T} - 0.0036884T$$

where T is expressed in Kelvin.[8] It can be identified and distinguished from other boron hydrides by its infrared spectrum,[9] mass spectrum,[10,11] or [11]B or [1]H n.m.r. spectrum.[12,13]

References

1. A. E. Stock, "Hydrides of Boron and Silicon," Cornell University Press, Ithaca, N. Y., 1933.
2. R. L. Hughes *et al.,* "Production of Boranes and Related Research," Academic Press, Inc., New York, 1967.
3. D. F. Gaines and R. Schaeffer, *Inorg. Chem.,* **3,** 438 (1963).
4. V. H. Miller and G. E. Ryschkewitsch, *J. Am. Chem. Soc.,* submitted for publication.
5. K. C. Nainan and G. E. Ryschkewitsch, *Inorg. Nucl. Chem. Letters,* **6,** 765 (1970).
6. D. F. Shriver, "The Manipulation of Air-sensitive Compounds," McGraw-Hill Book Company, New York, 1969.
7. R. T. Sanderson, "Vacuum Manipulation of Volatile Compounds," John Wiley & Sons, Inc., New York, 1948.
8. H. E. Wirth and E. D. Palmer, *J. Phys. Chem.,* **60,** 914 (1956).
9. L. V. McCarty, G. C. Smith, and R. S. McDonald, *Anal. Chem.,* **26,** 1027 (1954).
10. V. H. Diebeler *et al., J. Res. Natl. Bur. St.,* **43,** 97 (1949).
11. W. S. Koski, J. J. Kaufman, L. Friedman, and A. R. Irsa, *J. Chem. Phys.,* **24,** 221 (1956).
12. R. Schaeffer, J. N. Shoolery, and R. Jones, *J. Am. Chem. Soc.,* **79,** 4606 (1957).
13. G. R. Eaton and W. N. Lipscomb, "NMR Studies of Boron Hydrides and Related Compounds," p. 92, W. A. Benjamin, Inc., New York, 1969.

27. DIMETHYLAMINE-BORANE

$$2NaBH_4 + 2(CH_3)_2NH + I_2 \longrightarrow 2NaI + 2(CH_3)_2NH \cdot BH_3 + H_2$$

Submitted by K. C. NAINAN* and G. E. RYSCHKEWITSCH*
Checked by W. E. BRYANT,† G. BORN,† V. HALL,† and C. D. SCHMULBACH†

Amine-boranes are important boron-nitrogen compounds. They are prepared by various methods such as: direct combination of amine and diborane,[1,2] reaction of tetrahydroborates with ammonium salts,[3] transamination of amine-boranes,[4] or displacement of tetrahydrofuran from tetrahydrofuran-borane which is first prepared from boron trifluoride etherate and sodium tetrahydroborate,[5] or by less convenient methods like reduction of an appropriate

*Department of Chemistry, University of Florida, Gainesville, Fla. 32601.
†Department of Chemistry, Southern Illinois University, Carbondale, Ill. 62901.

boron compound using hydrogen or tetrahydroborate.[6-8]

The present method eliminates the intermediate preparation of diborane or ammonium salts, and does not require elevated temperatures, as in transamination. The concomitant risk of aminoborane formation is thus avoided. Finally, the present method generally produces high yields or pure product with ammonia, primary, secondary, or tertiary amines, diamines, and with phosphines from readily available starting materials in conventional apparatus.[9]

Procedure

■ **Caution.** *This synthesis should be carried out in a hood and away from flames, since the evolved hydrogen may cause an explosion or fire hazard.* In a tared 500-ml. two-necked flask containing 150 ml. of dry 1,2-dimethoxyethane (glyme)* gaseous dimethylamine (9.420 g., 0.209 mole) is dissolved by bubbling. Sodium tetrahydroborate (98%, metal hydrides, 8.957 g., 0.237 mole) and a stirring bar are added to the mixture. The main neck of the reaction flask is fitted with a 125-ml. pressure-compensating dropping funnel, containing 18.00 g. of iodine (0.142 g. atom) in 95 ml. of dry glyme, while the other neck is connected to a bubbler containing a benzene-amine mixture (any tertiary amine may be used; pyridine is satisfactory) to remove any gaseous boranes from the effluent hydrogen gas. The apparatus is purged with dry nitrogen gas for about 5 minutes before the addition of iodine. During a period of 2 hours, the iodine solution is added dropwise to the stirred reaction mixture while the evolved hydrogen gas is allowed to pass through the bubbler. Stirring of the reaction mixture is continued for another 15 minutes after the iodine addition is completed. The solvent is then evaporated in a stream of dry nitrogen or by vacuum evaporation in a rotary evaporator. The resulting solid residue is extracted with 140 ml. of benzene which has been stored over molecular sieve 3A, and the extract is filtered under a blanket of nitrogen. The residue is washed with 60 ml. of benzene in small portions. The filtrate with the washings is then evaporated to dryness in a stream of nitrogen, and the di-

---

*The solvent (Eastman) was dried by storing over molecular sieve 3A and CaH_2 for about a week. Alternatively it may be dried by distilling from CaH_2.

methylamine-borane obtained is dried further under vacuum for about 30 minutes.* The yield is 7.032 g. (84.4% based on iodine, m.p. 33°C.).

Further purification can be achieved by dissolving a weighed portion of the crude product (1.320 g.) in 15 ml. of dichloromethane and filtering. To the filtrate with 15-ml. washings, 150 ml. of *n*-hexane is added, and a slow stream of nitrogen gas is passed over the dichloromethane-hexane mixture. During the evaporation of the solvents, white crystalline borane is slowly precipitated, and when the final volume of the solvent mixture is reduced to about 45–50 ml., the borane is filtered and dried for about 10–15 minutes by aspirator suction. The compound is dried further by vacuum pumping for about 35 minutes. The yield is 1.004 g. (76%, m.p. 34.5–35°). One more recrystallization from dichloromethane-hexane gives a sample having m.p. 36°; lit. m.p. 36°.[10] The checkers found that dimethylamine-borane forms beautiful white needles on recrystallization from pure hexane.

Properties

Dimethylamine-borane is a white crystalline compound, soluble in many organic solvents, as are other amine-boranes. Unlike most borane adducts of tertiary amines and phosphines, it is also soluble in water. The vapor pressure of the adduct is 0.1 torr at 25°.[10] The compound is reasonably stable in neutral or basic aqueous solution at room temperature but decomposes in acidic solution. On heating, dimethylamine-borane loses hydrogen to form dimethylamino-borane.[11]

The infrared spectrum (Nujol and KBr) shows characteristic absorptions for coordinated BH_3 in the region 2260–2400 cm.$^{-1}$ for B—H stretch, and 1165 cm.$^{-1}$ for B—H bend. The proton n.m.r. spectrum in dichloromethane gives a doublet ($J = 6$ Hz.) for the methyl protons at 2.50 p.p.m. down field from tetramethylsilane (internal reference). The protons attached to the boron are observed as a symmetrical quartet, as expected for BH_3 adducts; three peaks

---

*Because of its volatility dimethylamine-borane may be partially lost during the vacuum-drying process. With less volatile adducts this problem is minimized.

are clearly visible but one peak is partially masked by the doublet because of the methyl protons. The center of the quartet appears at 1.45 p.p.m. down field from tetramethylsilane (internal reference), and the coupling constant (J_{B-H}) is 94 Hz. The ^{11}B spectrum of the compound in dichloromethane is a quartet (1:3:3:1), with chemical shift, $\delta = 32.4$ p.p.m. up field from trimethyl borate (external reference), and coupling constant $J_{B-H} = 94$ Hz.

References

1. K. Niedenzu and J. W. Dawson, "Boron-Nitrogen Compounds," Academic Press, Inc., New York, 1965.
2. R. A. Geanangel and S. G. Shore, in "Preparative Inorganic Reactions," W. L. Jolly (ed.), Vol. 3, p. 133, Interscience Publishers, a division of John Wiley & Sons, Inc., New York, 1966.
3. G. W. Shaeffer and E. R. Anderson, *J. Am. Chem. Soc.*, **71**, 2143 (1949).
4. R. A. Baldwin and R. M. Washburn, *J. Org. Chem.*, **26**, 3549 (1961).
5. R. C. Moore, S. S. White, Jr., and H. C. Keller, *Inorganic Syntheses*, **12**, 109 (1970).
6. R. Köster, *Angew. Chem.*, **69**, 64 (1957).
7. E. C. Ashby and W. E. Foster, *J. Am. Chem. Soc.*, **84**, 3407 (1962).
8. K. Lang and F. Schubert, U.S. Patent 3,037,985 (1962).
9. K. C. Nainan and G. E. Ryschkewitsch, *Inorg. Chem.*, **8**, 2671 (1969).
10. E. R. Alton, R. D. Brown, J. C. Carter, and R. C. Taylor, *J. Am. Chem. Soc.*, **81**, 3550 (1959).
11. G. E. Ryschkewitsch and J. W. Wiggins, *Inorg. Chem.*, **9**, 314 (1970).

28. DIMETHYLAMINE-TRIBROMOBORANE

$$(CH_3)_2NH \cdot BH_3 + 2Br_2 \longrightarrow (CH_3)_2NH \cdot BBr_3 + HBr + H_2$$

Submitted by W. H. MYERS* and G. E. RYSCHKEWITSCH*
Checked by W. E. BRYANT,† G. BORN,† V. HALL,† and C. D. SCHMULBACH†

The classical method of preparation of amine-trihaloboranes involves the direct addition of the amine to the boron trihalide.[1,2] However, when this method is used in an attempt to prepare boron trihalide adducts of secondary amines, one obtains a mixture of

*Department of Chemistry, University of Florida, Gainesville, Fla. 32601.
†Department of Chemistry, Southern Illinois University, Carbondale, Ill. 62901.

products,[3,4] and the isolation of the pure adduct from this mixture of products is difficult. The following synthesis has the advantage of starting with easily obtainable materials and achieving high yields of pure dimethylamine-tribromoborane under simple conditions. This synthesis is a specific example of a general procedure for the preparation of trihaloborane adducts of various amines and phosphines.[5]

Procedure

■ **Caution.** *This synthesis should be carried out in a hood and away from flames or spark sources, since the evolved hydrogen may cause an explosion or fire hazard and the evolved hydrogen bromide creates a corrosion and health hazard.* Into a 100-ml. two-necked flask containing a magnetic stirring bar are placed 2.95 g. (50.0 mmoles) of dimethylamine-borane and 40 ml. of dry dichloromethane. One neck of the flask is equipped with a 100-ml. pressure-equalizing dropping funnel; the other neck is fitted with an outlet tube equipped with a Nujol bubbler. Bromine (16.05 g., 100.4 mmoles) is dissolved in 40 ml. of dry dichloromethane, the solution is put into the dropping funnel, and the dropping funnel is stoppered. The bromine solution is added to the borane solution very slowly and with vigorous stirring. The time for the addition is at least 3 hours. (Faster addition will result in loss of too much hydrogen bromide, and thus incomplete reaction.) As the last of the bromine solution is added, a copious, crystalline white precipitate forms, and the solution, which was previously colorless, turns slightly yellow. (The precipitate is sometimes yellow.) All the volatile substances are removed under vacuum on a rotary evaporator. The white solid which remains is taken, still in the flask, into the dry-box, and the product is transferred into a tared bottle. Yield is 13.0 g. (88.0%, based on dimethylamine-borane). *Anal.* Calcd. for $C_2H_7NBBr_3$: C, 8.13; H, 2.39; N, 4.74; Br, 81.09. Found: C, 8.14; H, 2.36; N, 4.64; Br, 81.21.

Properties

Dimethylamine-tribromoborane is a white, crystalline solid, m.p. 150–152°. It is slightly soluble in benzene, dichloromethane, and carbon tetrachloride and is insoluble in petroleum ether. The compound may be precipitated from dichloromethane solution with petroleum ether to give moderate (72%) recovery of product. The compound fumes in air and is rapidly hydrolyzed in neutral water.

The infrared spectrum shows a strong, sharp band at 3150 cm.$^{-1}$, assigned to N—H stretch, and a broad band at 650–700 cm.$^{-1}$, assigned to coordinated BBr_3 vibrations.[6] The entire infrared spectrum is : 3220 (w), 3150 (s), 2710 (w), 2680 (w), 2480 (w), 1470 (m), 1460 (m), 1445 (m), 1435 (m), 1410 (m), 1370 (m), 1335 (m), 1220 (vw), 1200 (w), 1140 (m), 1130 (m), 1040 (w), 1000 (m), 895 (s), 825 (m), 790 (s), 710 (m), 680 (s), 665 (s), 455 (m).

The proton resonance spectrum, taken in dichloromethane solution, shows a multiplet at 3.03 p.p.m. down field from internal tetramethylsilane, which resembles a sextet, of 1:1:2:2:1:1 intensity. Under high resolution, eight peaks can be seen. Coupling constants were found to be: $J_{HNCH} = 5.4$ Hz.; $J_{BNCH} = 3.4$ Hz.

The ^{11}B resonance spectrum, taken at 32.1 MHz. in dichloromethane solution, shows a single peak at 24.8 p.p.m. up field from external trimethyl borate.

References

1. E. Wiberg and W. Sütterlin, *Z. Anorg. Allgem. Chem.,* **202,** 31 (1931).
2. R. C. Osthoff, C. A. Brown, and H. Clark, *J. Am. Chem. Soc.,* **73,** 4045 (1951).
3. W. Gerrard, H. R. Hudson, and E. F. Mooney, *J. Chem. Soc.,* **1960,** 5168.
4. R. A. Geanangel and S. G. Shore, in "Preparative Inorganic Reactions," W. L. Jolly (ed.), Vol. 3, p. 133, Interscience Publishers, a division of John Wiley & Sons, Inc., New York, 1966.
5. W. H. Myers, G. E. Ryschkewitsch, M. A. Mathur, and R. W. King, to be published.
6. R. L. Amster and R. C. Taylor, *Spectrochim. Acta,* **20,** 1487 (1964).

29. METHYLDIPHENYLPHOSPHINE-BORANE AND DIMETHYLPHENYLPHOSPHINE-BORANE

Submitted by M. A. MATHUR,* W. H. MYERS,* H. H. SISLER,* and G. E. RYSCHKEWITSCH*
Checked by G. E. BROWN,† P. KURIAN,† and C. D. SCHMULBACH†

Dimethylphenylphosphine-borane has been prepared in good yield by the displacement of trimethylamine by the less volatile dimethylphenylphosphine from trimethylamine-borane.[1] This method could be extended to the syntheses of other phosphine-boranes, but it involves the use of expensive trimethylamine-borane. The direct combination[2] of diborane with substituted phosphines could give good yields of pure phosphine-boranes, but the required use of vacuum-line techniques limits its utility to small-scale preparations. For the bulk synthesis of phosphine-boranes, the method reported by Nainan and Ryschkewitsch[3] has been found to be most suitable, and we have now extended this method to the syntheses of dimethylphenylphosphine-borane and methyldiphenylphosphine-borane.

The syntheses of these boranes are carried out in two steps: (1) the synthesis of the phosphine and (2) the conversion of phosphine into phosphine-borane. Since the steps involved in these two cases are essentially the same, only the synthesis of methyldiphenylphosphine-borane will be described here in detail. (A brief account of the synthesis of dimethylphenylphosphine-borane is given in Sec. B.)

A. METHYLDIPHENYLPHOSPHINE-BORANE

$$(C_6H_5)_2ClP + CH_3MgI \longrightarrow (C_6H_5)_2(CH_3)P + MgICl$$

$$(C_6H_5)_2(CH_3)P + NaBH_4 + \tfrac{1}{2}I_2 \longrightarrow$$

$$(C_6H_5)_2(CH_3)PBH_3 + \tfrac{1}{2}H_2 + NaI$$

Procedure

■ **Caution.** *Chlorodiphenylphosphine, dichlorophenylphosphine, methyldiphenylphosphine, and dimethylphenylphosphine are noxious,*

*Department of Chemistry, University of Florida, Gainesville, Fla. 32601.
†Department of Chemistry, Southern Illinois University, Carbondale, Ill. 62901.

toxic liquids. The chlorides are readily hydrolyzed, and all are oxidized in air. Consequently, they should be handled in a dry inert atmosphere. Glassware contaminated with the phosphines is cleaned by first immersing in a dilute solution (5.25%) of sodium hypochlorite to oxidize the phosphine. (A laundry bleach such as Clorox is satisfactory.)

1. Preparation of Methyldiphenylphosphine[4]

Magnesium metal (36.5 g., 1.50 g. atom) is placed in a 5-l. four-necked flask containing 1 l. of diethyl ether. The flask is equipped with a mercury-seal mechanical stirrer, a nitrogen inlet, a nitrogen outlet through a Dry Ice cold finger and a drying tube, and a 250-ml. pressure-equalizing dropping funnel. The contents are kept in the nitrogen atmosphere throughout the preparation. A solution of methyl iodide (93.4 ml., 1.50 mmoles) in 150 ml. of ether is added through the dropping funnel over a period of one hour. The resulting solution is stirred for 15 hours, then cooled in an ice bath for 2 hours. A solution of chlorodiphenylphosphine* (188.0 ml., 1.00 mole) in 50 ml. of ether is added slowly through the dropping funnel to the stirred solution of methyl magnesium iodide. The reaction is vigorous initially (the reaction is still more vigorous when dichlorophenylphosphine is used) but slows as the last half of the chlorodiphenylphosphine is added. After all the chlorodiphenylphosphine has been added, the resulting solution is refluxed for 2 hours and again cooled in an ice-salt bath. One liter of a saturated aqueous solution of ammonium chloride† is added slowly through the dropping funnel. The reaction mixture is stirred for 10 hours, then allowed to stand for 48 hours. The ethereal layer is separated from the aqueous layer by decantation. To the aqueous layer, 500 ml. of ether is added, and the mixture is stirred for 2 hours. The ethereal and aqueous layers are then separated. The two ethereal layers are mixed, and most of the ether is evaporated under reduced pressure; the remaining ether is removed on a rotary evaporator. The crude methyldiphenylphosphine is dried over anhydrous sodium sulfate

---

*Available from Ventron Corp., Beverly, Mass. 01915, and several other suppliers.

†The reaction with ammonium chloride is highly exothermic and may take place with the formation of a solid mass and rapid evaporation of ether. The checkers found that a smaller quantity (200 ml.) of NH_4Cl solution gave satisfactory results, but they did not observe phase separation.

and distilled under reduced pressure. The fraction boiling at 152–155° at 10.5 torr is collected. Yield is 167.7 g. (83.8%, based on chlorodiphenylphosphine).

When the scale of the reaction is reduced or when the time is shortened for settling of the two-layer system obtained after addition of saturated aqueous ammonium chloride solution to the ether solution, the yield is markedly reduced.

2. Conversion of Methyldiphenylphosphine to Methyldiphenylphosphine-Borane

■ **Caution.** *This synthesis should be carried out in a hood and away from flames or spark sources, since the evolved hydrogen may cause an explosion or fire hazard.*

A 1-l. three-necked flask is equipped with a mercury-seal mechanical stirrer, a pressure-equalizing dropping funnel, and an outlet for the reaction gases through a pyridine-benzene trap to destroy any gaseous boranes which may be formed. The vessel is flushed with nitrogen before and after adding the phosphine. Methyldiphenylphosphine (50.05 g., 0.250 mole) and finely ground sodium tetrahydroborate (11.5 g., 0.300 mole) are placed in the flask and covered with 500 ml. of anhydrous 1,2-dimethoxyethane. The dropping funnel is charged with iodine (33.3 g., 0.131 mole) dissolved in 200 ml. of 1,2-dimethoxyethane. The iodine solution is added dropwise over a one-hour period to the stirred suspension of sodium tetrahydroborate and methyldiphenylphosphine. The reaction is immediate, as indicated by the rapid disappearance of the iodine color. After all the iodine has been added, the solvent is removed by blowing N_2 over the solution. The slurry thus obtained is extracted with 750 ml. of benzene. The borane-benzene solution is filtered through a fine-porosity glass-frit filter. A large quantity of white solid (mainly NaI with a small amount of unreacted $NaBH_4$) is recovered and washed with two additional 20-ml. portions of benzene. (The filtration is very slow.) The benzene is evaporated from the solution under a good vacuum, leaving a viscous liquid weighing 51.7 g. The yield based on methyldiphenylphosphine is 96.6%. *Anal.* Calcd. for $C_{13}H_{16}BP$: C, 72.95; H, 7.53. Found: C, 72.88; H, 7.46.

Properties

Methyldiphenylphosphine-borane is a clear, colorless, viscous liquid which is stable on standing in air. It is readily soluble in dichloromethane, benzene, acetone, and chloroform, but it is insoluble in petroleum ether and water. It has a density of 1.042 g./cc. at 25.5°.

The infrared spectrum shows a strong band at 2370 cm.$^{-1}$, characteristic of BH_3 coordinated to a donor group, and a band at 1110 cm.$^{-1}$, characteristic of four-coordinate phosphorus. The entire infrared spectrum is given in Table I.

TABLE I   Infrared Data for Phosphine-Boranes (cm.$^{-1}$)

A. $(C_6H_5)_2(CH_3)PBH_3$:
3060 (w), 2920 (m), 2370 (s), 2250 (w), 1490 (w), 1440 (m), 1420 (w), 1335 (w), 1310 (w), 1295 (w), 1130 (w), 1110 (m), 1060 (s), 1000 (w), 890 (s), 775 (w), 735 (s), 690 (s), 590 (m), 490 (w), 460 (w), 420 (m)

B. $(C_6H_5)(CH_3)_2PBH_3$:
3060 (w), 2990 (m), 2920 (w), 2350 (s), 2320 (sh), 2250 (w), 1490 (m), 1440 (m), 1425 (m), 1340 (w), 1310 (w), 1290 (m), 1190 (w), 1130 (m), 1115 (m), 1060 (s), 1000 (w), 940 (m), 915 (s), 850 (m), 730 (s), 690 (s), 575 (m), 470 (m), 380 (m)

TABLE II   Nuclear Magnetic Resonance Data for Phosphine-Boranes

A. $(C_6H_5)_2(CH_3)PBH_3$
1. ^{1}H n.m.r. C_6H_5: $\delta = 7.50$ p.p.m.
   CH_3: $\delta = 1.83$ p.p.m. $J_{PCH} = 10.5$ Hz.
   BH_3: $\delta = 0.73$ p.p.m. $J_{BH} = 98$ Hz.
   $J_{PBH} = 16$ Hz.

2. ^{11}B n.m.r.: $\delta_B = 55.7 \pm 2$ p.p.m.
   $J_{BP} = 60 \pm 10$ Hz.
   $J_{BH} = 98 \pm 10$ Hz.

B. $(C_6H_5)(CH_3)_2PBH_3$
1. ^{1}H n.m.r. C_6H_5: $\delta = 7.50$ p.p.m.
   CH_3: $\delta = 1.54$ p.p.m. $J_{PCH} = 10.5$ Hz.
   BH_3: $\delta = 0.73$ p.p.m. $J_{BH} = 96$ Hz.
   $J_{PBH} = 16$ Hz.

2. ^{11}B n.m.r.: $\delta_B = 52.5 \pm 0.2$ p.p.m.
   $J_{BP} = 53 \pm 5$ Hz.
   $J_{BH} = 93 \pm 5$ Hz.

The proton resonance spectrum, taken in dichloromethane solution with tetramethylsilane as an internal standard, shows three sets of peaks: a complex multiplet in the aromatic region assigned to the phenyl protons, a doublet in the aliphatic region assigned to the methyl protons, and a weak, broadened quartet assigned to the borane protons. The phenyl to methyl intensity ratio is 10.0:3.3, which is consistent since the methyl doublet overlaps one of the four peaks owing to the borane protons. The chemical shifts and coupling constants are given in Table II.

The ^{11}B resonance spectrum, run at 25.2 MHz., in dichloromethane solution with trimethyl borate as an external standard, shows an octet of approximate intensity 1:1:3:3:3:3:1:1. Chemical shifts and coupling constants are given in Table II.

B. DIMETHYLPHENYLPHOSPHINE-BORANE

$$(C_6H_5)Cl_2P + 2CH_3MgI \longrightarrow (C_6H_5)(CH_3)_2P + 2MgICl$$

$$(C_6H_5)(CH_3)_2P + NaBH_4 + \tfrac{1}{2}I_2 \longrightarrow$$

$$(C_6H_5)(CH_3)_2PBH_3 + NaI + \tfrac{1}{2}H_2$$

Procedure

A solution of dichlorophenylphosphine (179.0 g., 1.00 mole) in 50 ml. of ether is added dropwise to a freshly prepared solution of methylmagnesium iodide (3.00 moles) using the procedure and equipment described in Sec. A. The resulting complex is decomposed with 1 l. of a saturated aqueous solution of ammonium chloride. The aqueous and ethereal layers are separated, and the crude dimethylphenylphosphine, obtained after the ether is evaporated, is dried over anhydrous sodium sulfate and fractionally distilled under reduced pressure. The portion which distills at 83–87° at 13–14 torr is collected. The yield is 69.3 g. (50.0%, based on dichlorophenylphosphine).

Dimethylphenylphosphine (35.7 ml., 0.250 mole) and finely ground sodium tetrahydroborate (11.5 g., 0.300 mole) are suspended in 300 ml. of anhydrous 1,2-dimethoxyethane. Iodine (33.3 g., 0.131 mole) in 200 ml. of 1,2-dimethoxyethane is added to this suspension through the dropping funnel. The amount of dimethylphenylphosphine-borane obtained after purification (similar to that discussed in Sec. A) is 34.9 g. (92.0% yield, based on dimethylphenylphosphine) *Anal.* Calcd. for $C_8H_{14}BP$: C, 63.22; H, 9.29. Found: C, 61.97; H, 9.70.

Properties

Dimethylphenylphosphine-borane is a clear colorless liquid with a foul odor. The compound is stable on standing in air. It is soluble in dichloromethane, chloroform, benzene, and acetone, but it is insoluble in petroleum ether and water.

The infrared spectrum shows a strong band at 2350 cm.$^{-1}$, characteristic of BH_3 coordinated to a donor group, and a band at 1115 cm.$^{-1}$, characteristic of tetrahedral phosphorus. The entire infrared spectrum is given in Table I.

The proton resonance spectrum, taken in dichloromethane solution with tetramethylsilane as an internal standard, shows three sets of peaks: a complex multiplet in the aromatic region assigned to the phenyl protons, a doublet in the aliphatic region assigned to the methyl protons, and a weak, broadened quartet assigned to the borane protons. The phenyl to methyl intensity ratio is 5.0:6.1, which is consistent, since the methyl doublet overlaps one of the four peaks owing to the borane protons. The chemical shifts and coupling constants are given in Table II.

The ^{11}B resonance spectrum, run at 25.2 MHz. on the neat liquid with trimethyl borate as an external standard, shows a pattern which resembles a quintet of 1:4:6:4:1 intensity. Low resolution caused a loss of fine structure. Chemical shift and coupling constants are given in Table II.

References

1. R. A. Baldwin and R. M. Washburn, *J. Org. Chem.,* **26,** 3549 (1961).
2. E. L. Gamble and P. Gilmont, *J. Am. Chem. Soc.,* **62,** 717 (1940).
3. K. C. Nainan and G. E. Ryschkewitsch, *Inorg. Chem.,* **8,** 2671 (1969).
4. G. M. Kosolapoff, "Organophosphorus Compounds," pp. 16–17, John Wiley & Sons, Inc., New York, 1950.

30. TRIPHENYLBORANE

$$BF_3 \cdot O(C_2H_5)_2 + 3C_6H_5MgBr \longrightarrow$$

$$B(C_6H_5)_3 + 3MgBrF + 3MgBrF + (C_2H_5)_2O$$

Submitted by R. KÖSTER,* P. BINGER,* and W. FENZL*
Checked by E. R. WONCHOBA† and G. W. PARSHALL†

Triphenylborane is best obtained from phenylmagnesium bromide and trifluoroborane-diethyl etherate,[1] although it is conveniently available on a small scale by pyrolysis of trimethylammonium tetraphenylborate.[2] According to the procedures described to date,[1,3] the Grignard reaction gives yields of 50–60%. However, triphenylborane is obtained in 80–90% yield if the following points are observed in the preparation:

1. The molar ratio of Grignard reagent to $BF_3 \cdot O(C_2H_5)_2$ must be exactly 3:1.

2. The Grignard solution is added dropwise to the $BF_3 \cdot O(C_2H_5)_2$ in such a rate that, until near the end of the reaction, the formation of a slightly soluble salt (tetraphenylborate) is largely excluded.

3. The addition of the Grignard solution and the distillation of the diethyl ether must be carried out without interruption until complete removal of the ether. In this way, precipitation of salts is avoided.

4. Biphenyl is best removed from the sublimed or distilled triphenylborane by recrystallization from heptane.

Procedure

■ **Caution.** *Triphenylborane is very air-sensitive; so all operations must be carried out in an inert atmosphere.*

The apparatus consists of a 4-l. three-necked flask with a paddle stirrer, a 1-l. dropping funnel, a simple still head fitted with a condenser, a 2-l. receiver, and an inert-gas inlet. The flask is charged

*Max Planck Institut für Kohlenforschung, 433 Mülheim/Ruhr, Germany.
†Central Research Department, E. I. du Pont de Nemours & Company, Wilmington, Del. 19898.

with 142 g. (1 mole) of trifluoroborane-diethyl etherate* in 1 l. of anhydrous xylene. An ethereal solution of 3 moles of phenyl-magnesium bromide† is added dropwise with stirring at 25–35° over a period of 3 hours. Only a little ether distills in this time. In the subsequent distillation of all the ether, it is important that the boiling point of xylene (138°) be attained at the end. The still hot solution is decanted from the precipitated magnesium salts through a bent glass tube into a 3-l. two-necked flask by pressure with an inert gas. (The checkers found it necessary to use a filter stick to prevent transport of the magnesium salts.) The magnesium salts are extracted twice with 500-ml. portions of hot xylene (120–130°). The combined xylene solutions are distilled under reduced pressure through a short, simple still head. After removal of the xylene, a *ca.* 5-g. intermediate fraction (mainly biphenyl) is obtained at b.p. <155° at 0.1 torr. Crude, nearly colorless triphenylborane distills in 225–g. yield at 155–166° at 0.1 torr. A single recrystallization from heptane under an inert atmosphere gives 217 g. (90%) of triphenylborane, m.p. 148°. *Anal.* Calcd. for $C_{18}H_{15}B$: C, 89.29; H, 6.24. Found: C, 89.06; H, 6.35.

The yellow-brown residue contains chiefly magnesium salts. The amount of the residue is greatly reduced if the ether is quantitatively removed before the xylene extraction.

Properties

Triphenylborane is a colorless, vacuum-sublimable, crystalline compound, which is very air- and moisture-sensitive but is thermally stable to *ca.* 380°. Because of a large molar melting-point depression, the melting point is a sensitive criterion of purity. In general crystals with a m.p. of 147–148° are obtained, but pure products melt at 151°.[5]

Triphenylborane must be handled and stored in an inert atmos-

---

*Available from BASF A. G., Ludwigshafen/Rhein, Germany, or Eastman Organic Chemicals, Rochester, N.Y.

†Prepared in about 90% yield from 3.3 g. atoms of freshly prepared magnesium turnings and 3.3 moles of bromobenzene (E. Merck, Darmstadt, Germany, or Eastman) in 1–1.5 l. of anhydrous diethyl ether.[4] The yield of Grignard reagent is conveniently determined by quantitative gas chromatographic analysis of the benzene formed by treatment of an aliquot with methanol.

phere. In moist air it slowly forms benzene and triphenylboroxine. Triphenylborane is soluble in toluene and xylene. It may be recrystallized from heptane almost without loss. Diethyl ether is not as good a solvent for the recrystallization of triphenylborane. The BC value of the compound can be readily determined by oxidation using the trimethylamine N-oxide method.[6]

References

1. E. Krause and R. Nitsche, *Chem. Ber.*, **55**, 1261 (1922).
2. K. A. Reynard, R. E. Sherman, H. D. Smith, and L. F. Hohnstedt, *Inorganic Syntheses*, **14**, 52 (1973).
3. H. C. Brown and V. H. Dodson, *J. Am. Chem. Soc.*, **79**, 2302 (1957).
4. G. S. Hiers, *Org. Syn., Coll.*, **1**, 550 (1941).
5. R. Köster, K. Reinert, and K. H. Müller, *Angew. Chem.*, **72**, 78 (1960).
6. R. Köster and Y. Morita, *Liebigs Ann. Chem.*, **704**, 70 (1967).

31. SODIUM TRIETHYLHYDROBORATE, SODIUM TETRAETHYLBORATE, AND SODIUM TRIETHYL-1-PROPYNYLBORATE

Submitted by P. BINGER* and R. KÖSTER*
Checked by E. R. WONCHOBA† and G. W. PARSHALL†

Trialkyl- and triarylboranes react with various sodium compounds, e.g., the hydride, cyanide, hydroxide, and amide as well as the alkyls, aryls, alcoholates, and phenolates, under mild conditions to form stable complex salts (borates) which contain four-coordinate boron as the central atom in the anion:[1-4]

$$NaX + BR_3 \longrightarrow Na[BXR_3]$$

X = H, CN, NH_2, OH, OR, alkyl, aryl, 1-alkenyl, 1-alkynyl

Sodium trialkylhydroborates obtained in this way are gentle, selective reagents for synthesis of transition-metal hydrides[5] and react with 1-alkynes to give trialkyl-1-alkynylborates.[4]

*Max Planck Institut für Kohlenforschung, 433 Mülheim/Ruhr, Germany.
†Central Research Department, Experimental Station, E. I. du Pont de Nemours & Company, Wilmington, Del. 19898.

The preparation of the salts through combination of the sodium compound and the trialkyl- or triarylborane generally proceeds smoothly. Only in a few instances are other procedures used. For example, sodium tetraalkylborates can be prepared from sodium, an alkyl chloride, and a trialkylborane.[6,7] In particular, sodium tetraethylborate is prepared most easily from sodium triethylhydroborate and ethylene.[7,8]

A. SODIUM TRIETHYLHYDROBORATE

$$NaH + B(C_2H_5)_3 \longrightarrow Na[BH(C_2H_5)_3]$$

At room temperature sodium hydride reacts exothermally with triethylborane in diethyl ether. The complex formation in hydrocarbons requires temperatures over $50°$. The preparation of sodium triethylhydroborate is best carried out in boiling benzene. The salt, which is readily soluble in benzene, is generally obtained in pure form after filtration of unreacted materials (e.g., sodium) and distillation of the solvent.

Procedure

■ **Caution.** *Triethylborane is spontaneously flammable, and sodium triethylhydroborate is easily hydrolyzed. The synthesis must be carried out in a protective gas (N_2, Ar) atmosphere with exclusion of air and moisture.*

A 1-l. three-necked flask equipped with dropping funnel, a precision-fit mechanical stirrer, and a reflux condenser (fitted with a gas bubbler and a tee fitting for admission of an inert gas) is charged with 50.4 g. (2.1 moles) of sodium hydride powder* in 200 ml. of sodium-dried benzene. To the refluxing ($80°$) mixture is added dropwise with strong stirring 196 g. (2.0 moles) of triethylborane† over 2 hours. After another 2 hours of heating under reflux, the insoluble materials (e.g., sodium metal) are filtered in an inert atmosphere, and the filtrate is freed of benzene by distillation. The

*Metal Hydrides, Inc., Beverly, Mass.

†Available from Alfa Inorganics, P.O. Box 159, Beverly, Mass. 01915, or prepared by the method of reference 9.

last traces of benzene are removed under vacuum (0.1 torr). The pale-yellow, viscous residue of 217 g. (89%) of sodium triethyl-hydroborate slowly crystallizes at $-15°$, m.p. 30°. *Anal.* Calcd. for $C_6H_{16}BNa$: Na, 18.81; hydride H, 0.817. Found: Na, 18.57; hydride H, 0.813.

The reaction of sodium hydride with trimethylborane under similar conditions gives sodium hydrotrimethylborate.

Properties

Sodium triethylhydroborate is very air- and moisture-sensitive and must be handled in an inert atmosphere. In completely pure form, it is a colorless crystalline solid, m.p. 30°. However, the salt is often obtained as a brown-black oil because of impurities in the sodium hydride.* Nevertheless, such products can be used without further purification in reactions such as preparation of sodium tetraethylborate.

B. SODIUM TETRAETHYLBORATE

$$Na[BH(C_2H_5)_3] + C_2H_4 \longrightarrow Na[B(C_2H_5)_4]$$

Sodium tetraethylborate is obtained from sodium triethylhydroborate 140–180° under ethylene pressure (65–100 atmospheres).[7,8] The addition of triethylborane[7] is desirable but not absolutely necessary.[8] (■ **Caution.** *The entire operation must be carried out in an inert gas atmosphere with strict exclusion of moisture.*)

Procedure

A 500-ml. steel autoclave (suitable for use at 500 atmospheres at 300°) with manometer is charged with 201 g. (1.65 moles) of sodium triethylhydroborate and pressured to 80 atmospheres with ethylene. The mixture is heated with shaking at 140–150° for 2 hours. The pressure falls from 100 to 30 atmospheres at 150°. After cooling to 20°, an additional 60 atmospheres of ethylene pressure is added, and the mixture is agitated a further 2 hours at 140–150°.

*For example, powder-form NaH from Feldmühle, Düsseldorf, Germany.

After cooling to 20°, the unreacted ethylene is vented and the autoclave is emptied by injection of two 300-ml. portions of hot (80–100°) toluene and release of the hot solution through a glass tube by means of nitrogen pressure. Colorless sodium tetraethylborate crystallizes from the combined solutions on cooling. After filtration and vacuum drying, 232.8 g. (94% yield) of the salt is obtained, m.p. 145°. *Anal.* Calcd. for $C_8H_{20}BNa$: C, 64.04; H, 13.44; B, 7.20; Na, 15.32. Found: C, 63.94; H, 13.23; B, 7.01; Na, 14.87.

The salt can also be obtained directly as a solid product by opening the autoclave under an inert atmosphere. The product is obtained as a hard, faintly yellow mass, m.p. 142°. Recrystallization from toluene gives white crystals, m.p. 145°.

Properties

Sodium tetraethylborate is a white, crystalline compound of m.p. 145°. The light-sensitive and highly hygroscopic salt is insoluble in aliphatic hydrocarbons and poorly soluble in cold benzene or toluene. It is readily soluble in diethyl ether, tetrahydrofuran, and diglyme. Aqueous solutions react distinctly alkaline and liberate 1 mole of ethane on acidification.

C. SODIUM TRIETHYL-1-PROPYNYLBORATE

$$Na[BH(C_2H_5)_3] + HC{\equiv}CCH_3 \longrightarrow Na[B(C_2H_5)_3(C{\equiv}CCH_3)] + H_2$$

Sodium triethylhydroborate reacts with propyne (methylacetylene) with hydrogen evolution even at room temperature.[4] In place of pure propyne, a hydrocarbon mixture containing *ca.* 30% propyne and *ca.* 30% allene* can also be used.[4] Only the propyne reacts.

Procedure

A 500-ml. three-necked flask is fitted with gas inlet tube, mechanical stirrer, and reflux condenser with a − 78° (hexane-CO_2) cold

---

*MAPP gas from Dow Chemical Company, Midland, Mich. Gas chromatography showed 8.1% propene, 20.2% propane, 28.9% propyne, 29.7% allene, 1.3% cyclopropane, 2% isobutane, and 9.7% of 1-butene and isobutene.

finger for condensation of propyne connected via a $-78°$ trap (C_3 hydrocarbons) to a gas-volume meter to measure the hydrogen evolved. The apparatus is dried at $100°$ and is filled with argon or nitrogen.

Propyne* is added at room temperature to a stirred solution of 122 g. (1 mole) of sodium triethylhydroborate in 150 ml. of sodium-dried benzene. Hydrogen evolution begins immediately with slight evolution of heat. The gas addition is regulated so that hydrogen evolution is continuous and practically no propyne reflux is observed (*ca.* 4.8 l. of C_3H_4 per hour). After 2 hours (*ca.* 16 g. of C_3H_4), the salt begins to precipitate. In 5 hours, 22.6 l. (S.T.P.) of gas is evolved, and the reaction is finished. The precipitate is filtered. After washing with a little pentane and drying under vacuum (12 torr), 142 g. (89%) of sodium triethyl-1-propynylborate of m.p. $90°$ is obtained.

When using MAPP gas, no cold finger ($-78°$) is used in order to avoid retention of the mixed gases. From 244 g. (2 moles) of sodium triethylhydroborate in 300 ml. of benzene and 280 g. of MAPP gas, the foregoing procedure gives 44.3 l. (S.T.P.) of hydrogen and 280 g. (88%) of sodium triethyl-1-propynylborate. In addition, the 500-ml. $-78°$ cold trap contains 165 g. of colorless liquid. [Gas chromatographic analysis: 13.4% propene, 33.6% propane, 39.2% allene, 2.5% isobutane, and 11.3% of 1-butene and isobutene.]

Properties

The white crystalline sodium triethyl-1-propynylborate of m.p. $90°$ is completely stable at room temperature under an inert gas (N_2 or Ar). The salt is sparingly soluble in pentane or in hexane, soluble in benzene, and easily soluble in diethyl ether or in tetrahydrofuran. The compound can be recrystallized from benzene or from cyclohexane. The salt is oxidized by air and is quantitatively cleaved by water with formation of propyne.

The proton n.m.r. spectrum in dimethyl sulfoxide-d_6 solution shows a singlet propynyl CH_3 at $\delta 1.53$, a triplet methyl signal at $\delta 0.55$ ($J = 7.2$ Hz.), and a quartet methylene signal at $\delta - 0.22$ p.p.m. in the appropriate intensity ratio of 1:3:2.

*Matheson Gas Products, East Rutherford, N.J. 07073.

References

1. W. Tochtermann, *Angew. Chem.*, **78**, 355 (1966); *Angew. Chem. Int. Ed.*, **5**, 351 (1966).
2. M. F. Lappert, in "The Chemistry of Boron and Its Compound," E. L. Muetterties (ed.), p. 443, John Wiley & Sons, Inc., New York, 1967.
3. M. A. Grassberger, "Organische Borverbindungen," Chemische Taschenbücher, Vol. 15, Verlag Chemie GmbH, Weinheim, Germany, 1971.
4. P. Binger, G. Benedikt, G. W. Rotermund, and R. Köster, *Liebigs Ann. Chem.*, **717**, 21 (1968).
5. H. Bönnemann, *Angew. Chem. Int. Ed.*, **9**, 736 (1970).
6. K. Ziegler and H. Hoberg, *Angew. Chem.*, **73**, 577 (1961).
7. Y. B. Honeycutt, Jr., and Y. M. Riddle, *J. Am. Chem. Soc.*, **83**, 369 (1961).
8. R. Köster, U.S. Patent 3,163,679 (1964); *Chem. Abstr.*, **62**, 11706c (1965).
9. R. Köster, *Liebigs Ann. Chem.*, **618**, 38 (1958).

32. TETRAALKYLDIBORANES AND 9-BORABICYCLO-[3. 3. 1] NONANE DIMER

Submitted by R. KÖSTER* and P. BINGER*
Checked by E. R. WONCHOBA† and G. W. PARSHALL†

In contrast to trialkylboranes BR_3, the mono- and dialkylboranes R_2BH and RBH_2 are dimers with structures related to diborane(6); i.e., they are alkyldiboranes. With a few exceptions (borolane dimers and 9-borabicyclo[3.3.1]nonane dimer) they may also be in equilibrium at room temperature with small amounts of monomer. Characteristically there is rapid R/H exchange. Hence alkyldiboranes are in most cases mixtures of boranes in various degrees of alkylation. "Tetraalkyldiboranes" are therefore mixtures of alkyldiboranes and trialkylboranes, and the mixtures have mean compositions corresponding to a tetraalkylated diborane:

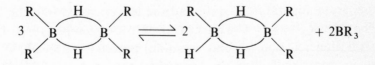

*Max Planck Institut für Kohlenforschung, 433 Mülheim/Ruhr, Germany.
†Central Research Department, E. I. du Pont de Nemours & Company, Wilmington, Del. 19898.

The chemistry of the alkyldiboranes is largely covered in recent review articles.[1-5] In practically all the preparative methods for organohydroboranes, the BH_3/BR_3 redistribution is important.

$$B_2H_6 + 4BR_3 \xrightarrow{(C_2H_5)_2O} 3R_4B_2H_2$$

Alkyldiboranes are most easily accessible through hydroboration of alkenes with diborane(6)[4,6] or through ligand exchange of metal hydrides such as $LiAlH_4$, NaH, $NaBH_4$, and Na $[BHR_3]$ with halo- and alkoxyboranes.[1,5,7] To obtain the stable alkyldiboranes such as borolane dimer[8] and 9-borabicyclo[3.3.1]nonane dimer,[8] the corresponding dienes are treated with alkyldiboranes:

$$B_2H_6 + 4BR_3 \longrightarrow 3R_4B_2H_2$$

Alkyldiboranes can also be obtained from trialkylboranes by hydrogenation under pressure at elevated temperatures.[9]

For preparation of the well-known[10,11] alkyldiboranes with ethyl, propyl, and butyl groups, it is convenient to react the available trialkylboranes[1,6] with diborane(6).

A. "TETRAETHYLDIBORANE" AND "TETRAPROPYLDIBORANE"

$$3NaBH_4 + 4BF_3 \cdot O(C_2H_5)_2 \longrightarrow 3NaBF_4 + 2B_2H_6 + 4(C_2H_5)_2O$$

$$4BR_3 + B_2H_6 \longrightarrow 3\text{"}R_4B_2H_2\text{"}$$

$$R = C_2H_5, C_3H_7$$

Diborane(6) is best generated in accord with the first equation in a separate apparatus (Fig. 4), and the gas is passed into the trialkylborane R_3B ($R = C_2H_5$, C_3H_7) at room temperature. The redistribution reaction (second equation) proceeds extremely sluggishly in pure trialkylborane. However, after addition of a little diethyl ether[12] or, better, some previously prepared alkyldiborane, diborane(6) reacts with heat evolution practically quantitatively with trialkylboranes.

■ **Caution.** *The characteristically disagreeable-smelling diborane is very toxic. Ethyl- and propyldiboranes are likewise foul-smelling and extremely sensitive to air and moisture. These spontaneously flammable compounds as well as triethyl- and tripropylboranes must be handled and stored under an inert gas such as pure nitrogen or pure argon.*

1. "Tetraethyldiborane"

■ **Note.** *See caution note above.*

Procedure

Xylene is purified by heating with metallic sodium and redistillation at atmospheric pressure (b.p. *ca.* 138°). Diglyme (diethylene glycol dimethyl ether or bis(2-methoxyethyl) ether) is purified by heating with sodium at 110° for *ca.* 3 hours (bare metal must still be visible) and distilling under vacuum (b.p. 57–58 at 12 torr). For grossly impure (OH impurities) diglyme, the procedure must be repeated. Trifluoroborane–diethyl ether complex* may be purified by distillation (b.p. 46° at 10 torr).

The apparatus of Fig. 4 is evacuated and filled with argon. In the 4-l. three-necked flask *A* (in an air bath) with dropping funnel *B* and a precision-fit metal-bladed stirrer *D*, 200 g. (5.26 moles) of sodium tetrahydroborate† powder is suspended in a mixture of 800 ml. of diglyme and 1.6 l. of xylene. In the 2-l. flask *G* with magnetic stirrer are placed 1015 g. (10.35 moles) of triethylborane and *ca.* 35 g. of an ethylated diborane (typically containing 21.1% hydridic H) or *ca.* 50 ml. of diethyl ether as discussed above. The dropping funnel *B* is charged with 1 kg. (7.0 moles) of trifluoroborane–diethyl etherate.

The addition of the first 3–5 ml. of the ether complex must be carried out especially carefully, with stirring, since gas evolution (chiefly hydrogen) may be extremely vigorous at first. The amount of the gas depends on the quality of the $NaBH_4$ and on the dryness

---

*Available from BASF AG, Ludwigshafen/Rhein, Germany, or from Eastman Organic Chemicals, Rochester, N.Y.

†Available from Bayer, A. G., Leverkusen, Germany, or from Metal Hydrides, Inc., Beverly, Mass.

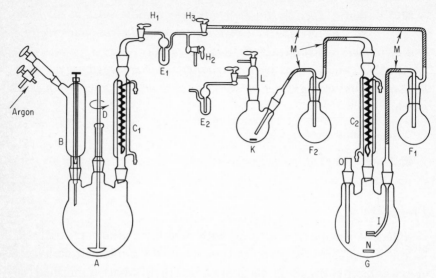

Fig. 4 Laboratory apparatus for the preparation of ethyl- and propyldiboranes. A, Four-liter three-necked flask; B, one-liter dropping funnel with pressure-equalizing arrangement; C1,C2, reflux condensers (condenser fluid is Aliphatin, a C_{10}-C_{13} paraffin mixture, b.p. ca. 190–230°, available from British Petroleum Corp.); D, KPG (precision-fit) stirrer with metal blades; E1,E2, bubble counter filled with Aliphatin; F1,F2, 250-ml. flasks with gas inlet tubes; G, two-liter three-necked flask; H, vacuum stopcocks; I, gas inlet tube with gas-dispersing frit; K, 250-ml. two necked flask; L, filling attachment; M, polyethylene tubing; N, magnetic stirrer; O, thermometer well.

of the suspending medium. Subsequently the trifluoroborane complex can be dropped in at a rate of 150 ml./hour. The mixture warms to about 35°, and diborane(6) is evolved continuously. It passes by way of the condenser $C1$, the bubbler $E1$, and 250-ml. trap $F1$ (cooled to $-78°$) through the gas-dispersion tube I into the triethylborane, which is stirred at room temperature. The almost quantitative reaction of diborane(6) with $(C_2H_5)_3B$ is accompanied by a temperature rise to *ca.* 60°.* Shortly before the completion of the addition (*ca.* 6 hours), flask A is warmed to about 50°, and then a slow argon stream is swept through the apparatus from B in order to complete the removal of the diborane. Finally, ethylated diborane in flask G can be transferred under an inert

---

*Excess diborane escapes through C_2 and F_2 and is trapped in flask K, which contains 100 ml. of triethylamine. Fluctuations in pressure are corrected by passing in argon from L to E_2.

gas into a suitable distillation apparatus (■ **Caution.** *Toxic and spontaneously flammable*), or a Vigreux distillation head may be substituted for the condenser *C*2 (■ **Caution**).

The product from *G*, if an ethyldiborane is used as the catalyst, is distilled rapidly at atmospheric pressure without removal of a forerun to give 1105 g. of colorless ethylated diborane (15.8% hydridic H), b.p. 96–112° at 760 torr. The yellow liquid residue weighs 12.1 g. If diethyl ether is used as the catalyst, distillation of a forerun of *ca.* 200 ml. boiling up to 94° is necessary in order to obtain completely ether-free product. The checkers carried out this reaction on a one-tenth scale using diethyl ether as the catalyst, and then used the initial product as catalyst for a second one-tenth-scale preparation. The second run gave 91.9 g. of ethylated diboranes, b.p. 105–108°. *Anal.* Calcd. for $(C_2H_5)_4B_2H_2$: 320 cc. hydrolytic H_2 per gram. Found: 368 cc./g. (S.T.P.).

In order to obtain "tetraethyldiborane," 100 g. of the ethylated diborane (15.8% hydridic H) is mixed with 10.5 g. of triethylborane to give 110.5 g. of "tetraethyldiborane" (14.3% hydridic H).

The $NaBF_4$ suspension in flask *A* may be hydrolyzed carefully with moist diglyme under an N_2 atmosphere. (■ **Caution.** *Toxic gases may be evolved.*)

Properties

"Tetraethyldiborane" (14.3% hydridic H) and ethylated diboranes with similar hydridic H content are clear, colorless, mobile, spontaneously flammable liquids with very unpleasant odor. (■ *Toxic*!) They consist of an equilibrium mixture of more highly ethylated boranes and diborane(6). Infrared absorption bands of the BH_2B bridge at 1567 cm.$^{-1}$ and the BH or BH_2 stretching at 2500 and 2570 cm.$^{-1}$ are spectroscopically characteristic and well suited for rapid identification. The characteristic B—H stretching band (2500 cm.$^{-1}$) of the dihydroborane constituent disappears only after addition of 6 moles of triethylborane per mole of "tetraethyldiborane."

A mixture of the gross composition of "tetraethyldiborane" (ν_{BH_2B} 1567 cm.$^{-1}$, ν_{BH} 2500 cm.$^{-1}$) boils at 109–111° at atmospheric pressure. "Triethyldiborane" (ν_{BH} 2500, 2570 cm.$^{-1}$, ν_{BH_2B} 1567 cm.$^{-1}$)

distills at 105–106°. After rapid distillation and cooling, the distillates spontaneously warm from room temperature to about 30°.

Ethyldiboranes must be stored under inert gas (argon, nitrogen). Only the B—H bonds react at room temperature with water. The B—H content of ethyldiboranes is determined simply and quantitatively by hydrolysis. The B—C content may be easily determined quantitatively by a variant of the trimethylamine N-oxide method.[13]

2. "Tetrapropyldiborane"

■ **Note.**   *See caution note above.*

Procedure

Xylene, diglyme, and trifluoroborane–diethyl etherate are purified as in Sec. A1. Tripropylborane of b.p. 159°[11] used in this preparation consists of a mixture of 93 % tripropylborane and *ca.* 7 % isopropyl-dipropylborane.[14]

The preparation is analogous to that of "tetraethyldiborane" and employs the apparatus of Fig. 4. Flask *A* is charged with a suspension of 150 g. (3.95 moles) of sodium tetrahydroborate powder in 600 ml. of diglyme and 1.2 l. of anhydrous xylene and flask *G* with 1090 g. (7.78 moles) of tripropylborane and *ca.* 50 ml. of diethyl ether. Trifluoroborane–diethyl ether complex (750 g., 5.28 moles) in funnel *B* is added dropwise with stirring over a period of *ca.* 5 hours. (■ **Caution.** *Vigorous gas evolution occurs during addition of first 5 ml.)* The mixture in *G* warms to a maximum of 50°, and the ether boils under reflux. After completion of diborane evolution and purging with argon as in Sec. A1, the crude propyldiborane (1120 g.) is distilled through a 40-cm. column packed with 3-mm. glass helices. The first 45 ml. of diethyl ether is distilled at atmospheric pressure. Then, under vacuum, an intermediate fraction of *ca.* 75 g. distills up to 55° at 12 torr. Propyldiborane (960 g., 11.3 % hydridic H) then distills at 55–63° at 12 torr, leaving *ca.* 40 g. of yellow residue.

"Tetrapropyldiborane" (10.2 % hydridic H) is obtained by addition of 11 g. of tripropylborane to 100 g. of propyldiborane (11.3 % hydridic H). In order to obtain a propyldiborane composition without significant terminal B—H content, 80 g. of tripropylborane is added to *ca.* 20 g. of propyldiborane (11.3 % hydridic H).

Properties

The colorless, mobile, unpleasant-smelling "tetrapropyldiborane" of b.p. 58–63° at 13 torr, like the analogous "tetraethyldiborane," exists in equilibrium with various higher propylated boranes at room temperature. The mixture is practically undecomposed by rapid distillation at 155–162°. Propyldiboranes are characterized by infrared bands assignable to v_{BH_2B} at 1560–1565 cm.$^{-1}$ and v_{BH} at 2500 and 2570–2580 cm.$^{-1}$. They must be stored and handled under an inert gas (argon or nitrogen). The B—H content is best determined by measurement of the hydrogen evolution of hydrolysis at room temperature. The B—C content of the propyldiboranes, in combination with the B—H value, is easily determined by a variant of the trimethylamine N-oxide method.[13]

B. 9-BORABICYCLO[3. 3. 1]NONANE DIMER

$$2C_8H_{12} + 3(C_2H_5)_4B_2H_2 \longrightarrow (C_8H_{14}BH)_2 + 4B(C_2H_5)_3$$

Procedure

■ **Caution.** *The ethyldiboranes are spontaneously flammable and toxic! The reaction should be carried out in an efficient hood, and the boron reagents and products should be handled under an inert gas.*

The apparatus is a 1-l. three-necked flask equipped with a 500-ml. dropping funnel, magnetic stirrer, internal thermometer, and a distillation head with descending condenser and 500-ml. receiver. The exit from the receiver is fitted with an inert-gas bypass to ensure an inert atmosphere whenever the apparatus is opened. The apparatus is flushed with argon or nitrogen, and the flask is charged with 113 g. (1.05 moles) of 1,5-cyclooctadiene.* The diene is stirred while 216 g. of ethyldiborane (prepared in Sec. A, 15.3% hydridic H as determined by volumetric H_2 evolution on hydrolysis, 3.3 moles BH) is added dropwise over *ca.* 3 hours. The temperature in the mixture rises from *ca.* 20° to *ca.* 65°. Finally the mixture is heated

---

*Available from Chemische Werke Hüls AG, Marl, Germany, or from Phillips Petroleum Co., Bartlesville, Okla.

at 120–140° (internal temperature) for about 3.5 hours. Approximately 200 ml. of triethylborane distills along with a little tetraethyldiborane. (The measured B—H content is less than 0.1 %.) On slow cooling to room temperature, well-formed needles crystallize. After recrystallization from *ca.* 200 ml. of anhydrous heptane, 120 g. (94%) of bis(9-borabicyclo[3.3.1]nonane) (m.p. 148°) is obtained. *Anal.* Calcd. for $C_{16}H_{30}B_2$: C, 78.75; H, 12.39. Found: C, 78.48; H, 12.38.

Properties

The pure crystalline bis(9-borabicyclo[3.3.1]nonane) melts at 148°[15] and has a b.p. at 12 torr of 195°.[16] The compound, in contrast to the simple tetraalkyldiboranes, has a definite structure. Its infrared spectrum shows v_{BH_2B} at 1567 cm.$^{-1}$. The completely pure compound is stable at room temperature even with air access for long periods. However, storage and handling of the compound should be carried out in an inert-gas atmosphere. Water and alcohols react with it even at room temperature, with evolution of hydrogen. Alcoholysis is well suited for determination of the B—H content. The quantitative BC determination proceeds very well by a variant of the trimethylamine N-oxide oxidation method.[13]

Bis(9-borabicyclo[3.3.1]nonane) is a useful reagent for synthesis of aldehydes by carbonylation of olefins.[17]

References

1. M. Grassberger, "Organische Borverbindungen," Chemische Taschenbücher, Vol. 15, Verlag Chemie GmbH, Weinheim, Germany, 1971.
2. H. D. Johnson and S. G. Shore, *Fortschr. Chem. Forsch.,* **15**, 87 (1970).
3. T. Onak, *Advan. Organometallic Chem.,* **3**, 263 (1966).
4. (a) H. C. Brown, "Hydroboration," W. A. Benjamin, Inc., New York, 1962; (b) G. Zweifel and H. C. Brown, *Org. Reactions,* **13**, 1 (1963).
5. M. F. Lappert, in "The Chemistry of Boron and Its Compounds," E. L. Muetterties (ed.), John Wiley & Sons, Inc., New York, 1967.
6. R. Köster, G. Griasnow, W. Larbig, and P. Binger, *Liebigs Ann. Chem.,* **672**, 1 (1964).
7. R. Köster, *ibid.,* **618**, 38 (1958).
8. R. Köster, *Angew. Chem.,* **72**, 626 (1960).
9. R. Köster, G. Bruno, and P. Binger, *Liebigs Ann. Chem.,* **644**, 1 (1961).
10. H. I. Schlesinger and A. O. Walker, *J. Am. Chem. Soc.,* **57**, 621 (1935).

11. H. I. Schlesinger, L. Horvitz, and A. B. Burg, *ibid.*, **58**, 407 (1936).
12. H. C. Brown and B. C. Subba Rao, *J. Org. Chem.*, **22**, 1135 (1957).
13. R. Köster and Y. Morita, *Liebigs Ann. Chem.*, **704**, 78 (1967).
14. G. Schomburg, R. Köster, and D. Henneberg, *Mitteilungsblatt der Chem. Ges. DDR,* Sonderheft 1960; *cf. Angew Chem.*, **75**, 1081 (1963).
15. R. Köster and M. A. Grassberger, *Liebigs Ann. Chem.*, **719**, 169, 185 (1968).
16. E. F. Knight and H. C. Brown, *J. Am. Chem. Soc.*, **90**, 5280 (1968).
17. M. Fieser and L. F. Fieser, "Reagents for Organic Synthesis," Vol. 3, p. 24, John Wiley & Sons, Inc., New York, 1972.

33. CHLORODIETHYLBORANE AND CHLORODIPHENYLBORANE

$$2BR_3 + BCl_3 \xrightarrow{R_2BH} 3BClR_2$$

Submitted by R. KÖSTER* and P. BINGER*
Checked by E. R. WONCHOBA† and G. W. PARSHALL†

Dialkyl- and diarylhaloboranes are accessible in various ways: (1) through haloboration of hydrocarbons; (2) through halogenation of organic-substituted diboranes and trialkylboranes; (3) through reaction of haloboranes with organometallic compounds (transmetallation).[1-4] An especially good procedure, illustrated below for chlorodiethylborane and chlorodiphenylborane, is the catalytic redistribution between organoboranes and trihaloboranes.[5]

Trialkyl- and triarylboranes react in the presence of certain B—H compounds,[5] usually even at room temperature and without isomerization of alkyl groups,[6] to give mono- and dihaloboranes BXR_2 and BX_2R (R = alkyl, aryl; X = F, Cl, Br). In this way, many fluoro-, chloro-, and bromoboranes are obtained, simply and in good yield, as well as many B-halo boron heterocycles.[5] Based on the molar ratio $BX_3:BR_3$, one obtains BRX_2 or BR_2X. With bulky organic groups, the reaction proceeds slowly or not at all. For example, trichloroborane reacts with triisopropylborane much more slowly than with tri-*n*-propylborane. The redistribution

*Max Planck Institut für Kohlenforschung, 433 Mülheim/Ruhr, Germany.
†Central Research Department, Experimental Station, E. I. du Pont de Nemours & Company, Wilmington, Del. 19898.

between tribromoborane and triisopropylborane is not catalyzed by alkyldiboranes even at 50–60°. Redistributions with triiodoborane have not yet been studied. Dialkyliodoboranes are, however, easily available from iodine and tetraalkyldiboranes.[5]

Both diborane and alkyldiboranes serve as B—H catalysts for the redistribution reaction. Very strongly associated B—H dimer compounds such as bisborolanes are not effective catalysts for ligand exchange below 100°.

General Procedure

Most dialkylhaloboranes as well as the trialkylboranes are liquid and are oxidized rapidly by atmospheric oxygen. (■ **Caution.** *Boranes with high vapor pressures such as chlorodiethylborane are spontaneously flammable.*) The B-haloboranes are easily hydrolyzed. Therefore, in all experiments, air and moisture must be excluded by operation under an inert gas (N_2, Ar).

The exothermic redistribution occurs only in the presence of a B—H compound. The mixture remains colorless. Trichloro- and tribromoborane also dissolve in trialkylboranes in the absence of B—H compounds with heat evolution, but generally these solutions are dirty yellow to brown. This discoloration must always be avoided by well-timed addition of the B—H catalyst. (■ *Add with care!*) In the preparation of fluoroboranes, this difficulty is avoided, since trifluoroborane or the more commonly used $BF_3 \cdot OR_2$ is only slightly soluble in trialkylboranes.

In the event that the catalyst is rendered inactive by protolysis (for example, by HX) in the course of an experiment (as evidenced by a colored solution), most of the dissolved trihaloborane must be removed by distillation. After addition of fresh alkyldiborane, the color of the mixture lightens or disappears and the experiment may be continued by addition of trihaloborane. Under no circumstances should the B—H catalyst be added to the BR_3/BX_3 (X = Cl, Br) mixture, because a sudden, violent reaction can occur. (■ *Fire hazard!*) Before distillation of the dialkylhaloborane, one destroys the catalyst by addition of a suitable olefin in order to avoid reversal of the synthesis reaction. In the preparation of diethylfluoroborane from triethylborane and dibutyl ether–trifluoroborane, catalyst destruction can be omitted because of the low boiling points of the products.

A. CHLORODIETHYLBORANE

$$2 \, B(C_2H_5)_3 + BCl_3 \xrightarrow{R_2BH} 3 \, BCl(C_2H_5)_2$$

Procedure (See also General Procedure)

A 500-ml. two-necked flask is fitted with a magnetic stirrer, thermometer, inlet tube, and water-cooled reflux condenser capped with a nitrogen bypass. The inlet tube, which extends beneath the surface of the liquid, is connected via a wash bottle and a T piece fitted with a three-way stopcock (for admission of an inert-gas diluent) to a 250-ml. supply cylinder of trichloroborane. The apparatus is dried by heating, is evacuated, and is filled with nitrogen. In order to produce a continuous gas stream, trichloroborane is diluted with nitrogen via the three-way stopcock.

To 147 g. (1.5 mole) of triethylborane,*[7] in the reaction flask is added 2 ml. of tetraethyldiborane.[8] At − 78°, 90 ml. (0.75 mole) of trichloroborane† is condensed in the supply cylinder. The supply cylinder is connected to the inlet tube, and trichloroborane vaporized by removal of the cooling bath is bubbled into the triethylborane at a rate such that the mixture warms to 40–60° (3–4 hours). Finally the nearly colorless mixture is allowed to cool to room temperature, and the inlet tube is replaced by a stopcock. 1-Octene (5 ml.) is added through it, and the mixture is stirred 1 hour at room temperature. (Instead of the 1-octene addition, ethylene can be bubbled through the mixture at room temperature for 1 hour.) The reflux condenser is replaced by a 30-cm. column packed with 3-mm. glass helices, and the mixture is distilled. (■ **Caution.** *A solvent-resistant grease such as Kel-F grease, available from the 3M Co., is required to resist the hot vapor.*) After a few milliliters forerun (b.p. 30–80°), one obtains 230 g. (98% yield) of completely colorless chlorodiethylborane, b.p. 81°.

---

*Available from Alfa Inorganics, P.O. Box 159, Beverly, Mass. 01915.
†Available from Elektroschmelzwerk Kempten GmbH, München, Germany, or Matheson Gas Products, East Rutherford, N.J. 07073.

The pyrophoric residue in the distillation flask may be deactivated by addition of benzene through the condenser under a nitrogen stream, followed by cautious addition of methanol.

By a similar procedure,[5] dibromopropylborane (b.p. 54–56° at 70 torr) may be prepared in 78% yield from tripropylborane[7,9] (0.1 mole) and tribromoborane (0.2 mole) with tetrapropyldiborane[8] (0.5 ml.) as the catalyst. Diethylfluoroborane (b.p. 39°) is similarly prepared[5] in 68% yield from triethylborane (1.25 moles) and dibutyl ether–trifluoroborane (0.54 mole) with tetraethyldiborane[8] (1.5 ml.) as the catalyst.

Properties

Chlorodiethylborane is a colorless, mobile, thermally stable liquid (b.p. 81°) which fumes strongly in air and is even pyrophoric. The B—Cl bond is very easily cleaved by hydrolysis. The compound must be handled and stored under an inert gas. For prolonged storage, it is desirable to store the substance in a sealed glass ampul.

B. CHLORODIPHENYLBORANE

$$(C_6H_5)_3B + C_6H_5BCl_2 \longrightarrow 2(C_6H_5)_2BCl$$

Procedure

■ **Caution.** *Phenylboranes are very sensitive to moist air. The starting materials, reaction mixtures, and products must be kept under an inert gas at all times.*

A 500-ml. two-necked flask is fitted with a magnetic stirrer and a still head with receiver. The flask is charged with 107 g. (0.443 mole) of triphenylborane[10] and 72 g. (0.45 mole) of dichlorophenylborane.[4] After addition of 1 ml. of tetraethyldiborane,[8] the mixture is heated with stirring at 40–45° for 4 hours. Most of the triphenylborane dissolves.

Ethylene is then bubbled through the solution for 1 hour at room temperature. Distillation under reduced pressure gives 10.3 g. of dichlorophenylborane (b.p. 30–70° at 0.2 torr), and 150 g. (85%) of chlorodiphenylborane (b.p. 101–102° at 0.2 torr). After cooling, it crystallizes as large platelets (m.p. 32°). *Anal.* Calcd. for $C_{12}H_{10}BCl$: C, 71.9; H, 5.03. Found: C, 71.8; H, 5.14. Triphenylborane remains in the flask, recovery 15.9 g.

The reaction of trichloroborane (15 g.) and triphenylborane (55.3 g.) in benzene or toluene at 25–45° with addition of 1 ml. of tetraethyldiborane[8] also gives chlorodiphenylborane in 65–75% yield.[5] Dichlorophenylborane is formed also.

Properties

Pure chlorodiphenylborane is a colorless crystalline compound of m.p. 32°[5] and of b.p. 271–272°.[1] The compound should be handled under an inert gas. The purification is best carried out by distillation under reduced pressure (b.p. 101–102°/0.2 torr).

References

1. K. Niedenzu, *Organometallic Chem. Rev.,* **1,** 305 (1966).
2. M. F. Lappert, in "The Chemistry of Boron and Its Compounds," E. L. Muetterties (ed.), p. 443, John Wiley & Sons, Inc., New York, 1967.
3. M. A. Grassberger, "Organische Borverbindungen," Chemische Taschenbücher, Vol. 15, Verlag Chemie GmbH, Weinheim, Germany, 1971.
4. P. M. Treichel, J. Benedict, and R. G. Haines, *Inorganic Syntheses,* **13,** 32 (1972).
5. R. Köster and M. A. Grassberger, *Liebigs Ann. Chem.,* **719,** 169 (1968), and references cited therein.
6. P. A. McCusker, G. F. Hennion, and E. C. Ashby, *J. Am. Chem. Soc.,* **79,** 5192 (1957).
7. R. Köster, *Liebigs Ann. Chem.,* **618,** 38 (1958).
8. R. Köster and P. Binger, *Inorganic Syntheses,* **15,** 142 (1974).
9. G. Schomburg, R. Köster, and D. Henneberg, *Z. Anal. Chem.,* **170,** 285 (1959).
10. R. Köster, P. Binger, and W. Fenzl, *Inorganic Syntheses,* **15,** 134 (1974); K. A. Reynard, R. E. Sherman, H. D. Smith, and L. F. Hohnstedt, *ibid.,* **14,** 52 (1973).

Chapter Six

GERMANIUM HYDRIDE DERIVATIVES

CHARLES VAN DYKE*

One area of contemporary inorganic chemistry that has undergone rapid development in recent years is the synthesis and study of the germanium hydrides and their derivatives. This research has been stimulated largely by the novel findings which have arisen from systematic studies of the hydrides of silicon, germanium's upper neighbor in the periodic table. The germanium compounds merit special attention because, with their availability, many of the well-known comparisons made between simple carbon and silicon compounds can now be extended to include the analogous germanium compounds. The results should be of great significance in assessing the bonding and chemical characteristics of the first-, second-, and third-row main group elements, particularly with regard to involvement of vacant d-orbitals in chemical bonding. This controversial facet of Group IV chemistry has been the subject of numerous investigations, mostly, however, with regard to silicon.

Thermal instability and lack of convenient preparative procedures have hindered studies of derivatives of the germanium hydrides. However, a great deal of progress has been made, and some very interesting studies on the physical and structural properties of the compounds have appeared in the literature. Preparative details have remained rather sketchy, and the main goal of this chapter is to provide specific instructions for the synthesis of certain represen-

*Carnegie-Mellon University, Pittsburgh, Pa. 15213.

tative derivatives of germane and digermane. It is hoped that new research efforts in germanium chemistry will be stimulated by stressing the recent synthetic achievements. Most of the compounds described are useful as intermediates in the synthesis of other germyl GeH_3-) or digermanyl (Ge_2H_5-) derivatives. Procedures for certain other compounds are included in order to illustrate general synthetic methods. The preparation of GeH_3K, an important intermediate in the synthesis of a variety of germyl compounds,[1] is not included. Its preparation by the deprotonation of GeH_4 has been described in an earlier volume of this series.[2] A review of the important procedures used in the synthesis of derivatives of germane and digermane has recently been given in the literature.[3]

Germane, the starting material from which most of the derivatives of the germanium hydrides are prepared, is now commercially available from Matheson Gas Products. It is also conveniently prepared by the reduction of GeO_2 with $NaBH_4$. Details for the synthesis of GeH_4, Ge_2H_6, and Ge_3H_8 by this method have been given in Volume VII of this series.[4] An alternate, related procedure which is particularly good for the synthesis of germane has also been described.[5,6] Details for preparing reasonably large quantities of the higher germanes by subjecting GeH_4 to a silent electric discharge are available.[7] It is strongly recommended that all synthetic work with the germanium hydrides and their derivatives be carried out in high-vacuum systems of a conventional design.[8]

References

1. R. M. Dreyfuss and W. L. Jolly, *Inorg. Chem.*, **10**, 2567 (1971).
2. D. S. Rustad, T. Birchall, and W. L. Jolly, *Inorganic Syntheses*, **11**, 128 (1968).
3. C. H. Van Dyke, in "Preparative Inorganic Reactions," W. L. Jolly (ed.), Vol. 6, p. 157, Interscience Publishers, a division of John Wiley & Sons, Inc., New York, 1971.
4. W. L. Jolly and J. E. Drake, *Inorganic Syntheses*, **7**, 34 (1963).
5. T. S. Piper and M. K. Wilson, *J. Inorg. Nucl. Chem.*, **4**, 22 (1957).
6. J. E. Griffiths, T. N. Srivastava, and M. Onyszchuk, *Can. J. Chem.*, **40**, 579 (1962).
7. S. D. Gokhale, J. E. Drake, and W. L. Jolly, *J. Inorg. Nucl. Chem.*, **27**, 1911 (1965).
8. D. F. Shriver, "The Manipulation of Air-sensitive Compounds," McGraw-Hill Book Company, New York, 1969.

34. BROMOGERMANE AND DIBROMOGERMANE

$$Br_2 + GeH_4 \longrightarrow GeH_3Br + HBr$$

$$2Br_2 + GeH_4 \longrightarrow GeH_2Br_2 + 2HBr$$

Submitted by M. F. SWINIARSKI* and M. ONYSZCHUK*
Checked by M. A. FINCH† and C. H. VAN DYKE†

Bromo- and dibromogermane are useful starting reagents for the preparation of a variety of derivatives containing germyl (GeH$_3$-) and germylene (>GeH$_2$) groups.[1] They were first prepared, in low yields, by the reaction of germane with hydrogen bromide in the presence of aluminum bromide as catalyst.[2] Other methods have involved: passing germane over heated silver bromide,[3] the exchange reaction of chlorogermane with hydrogen bromide,[4] and the reaction of chlorogermane with boron tribromide.[5] The highest yields are obtained conveniently by the direct reaction of bromine with germane [6] according to the following procedure patterned after the preparation of bromosilane by the controlled bromination of silane.[7] The synthesis, including purification and purity checks, can be completed in about 10 hours.

Procedure

■ **Caution.** *All operations should be carried out in a well-ventilated area because of the toxicity of bromine and of germane derivatives.*

The assembly (Fig. 5) is attached to a conventional vacuum line equipped with a distillation train.[8-10] All stopcocks and ground-glass joints are lubricated with an inert, self-thickened, halocarbon grease,‡ which is unaffected by bromine. Germane§ is best prepared

*McGill University, Montreal, Quebec, Canada.
†Department of Chemistry, Carnegie-Mellon University, Pittsburgh, Pa. 15213.
‡Available from Halocarbon Products Corp., 82 Burlews Court, Hackensack, N.J., Series 25–10M. Apiezon N stopcock grease is attacked by bromine but was successfully used by the checkers. However, it must be replaced before the vacuum line is used for other work.
§Available from Matheson Gas Products, East Rutherford, N.J. 07073.

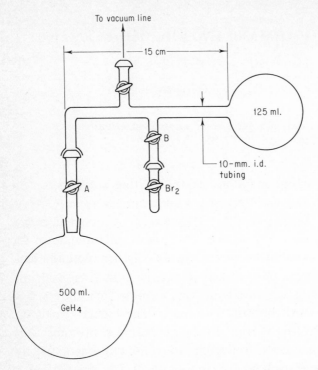

Fig. 5. Vacuum-line accessory for bromination of germane.

by the reduction of germanium dioxide with sodium tetrahydroborate.[6,8] Care should be taken to ensure that the total amount of germane used in the synthesis does not produce more than atmospheric pressure upon expansion in the reaction vessel.

In a typical preparation 18.8 mmoles of germane is condensed into the 500-ml. flask (Fig. 5) cooled by liquid nitrogen to $-196°$. A slightly less than equimolar amount of bromine, 17.9 mmoles, is placed in the small tube, attached to the vacuum apparatus, and degassed. With stopcock A and the stopcock to the vacuum line closed, stopcock B is opened and bromine is allowed to expand into the 125-ml. bulb. Stopcock B is then closed and the bromine vapor is condensed into the 500-ml. reaction vessel. (■ **Caution.** *An explosion may occur if too much bromine is added at one time!*) The reaction vessel should contain just enough bromine to be a moderate yellow-orange color. Stopcock A is closed and the liquid nitrogen is removed. The mixture of germane and bromine is allowed to warm gradually until the color of bromine just disappears, and the reaction vessel is quickly cooled again with liquid nitrogen. Bromination normally proceeds slowly under such mild conditions, but occasionally a vigorous reaction occurs as shown by the sudden formation of a cloud in the reaction vessel. The procedure is repeated 10 to 15 times until about half the bromine is consumed. Approximately 2 hours are required to reach this point.

At this stage the product mixture and unconsumed germane are distilled through a trap held at $-95°$ (toluene slush) into a liquid-nitrogen trap, without pumping. Only germane and hydrogen bromide pass through to be condensed at $-196°$ while the product mixture, consisting of bromo-, dibromo-, tribromo-, and tetra-bromogermane, is retained in the $-95°$ trap. The recovered mixture of germane and hydrogen bromide is condensed back into the 500-ml. reaction vessel cooled to 196°. The second half of the bromine is now added in the same manner as the first half.

The products of the two halves of the synthesis are condensed together and separated into their components by distillation through a series of interconnected traps kept at $0°$, $-45°$, $-95°$, $-160°$, and $-196°$ in this order and without pumping. Tribromogermane and tetrabromogermane are retained at $0°$, dibromogermane (4.8 mmoles, 46%) at $-45°$, bromogermane (5.6 mmoles, 54%) at $-95°$, hydrogen bromide at $-160°$, and germane (8.4 mmoles) at $-196°$. The dibromo- and bromogermane fractions are redistilled through the same distillation train until vapor pressure, molecular weight, and infrared measurements confirm their purities.

Yields of bromogermane in this synthesis are usually in the range of 50 to 65%, and occasionally as high as 88–91% based on the amount of germane consumed. More frequent distillations during the bromine-addition operation (e.g., after addition of five small portions of bromine) may improve the yield of the monobromo derivative. Attempts to improve the yield of bromogermane by varying the mole ratio of germane to bromine in the over-all reaction have been unsuccessful. Increased yields of dibromogermane result when a higher ratio of bromine to germane is used.[11]

Smaller quantities (about 1 mmole) of the compounds can be made by combining in one step approximately 2-mmole quantities of germane and bromine, using the apparatus previously described. The yields are approximately the same as in the large-scale synthesis. (■ **Caution.** *If larger amounts of reactants are combined* *directly, there is a danger of an explosion due to the vigor of the bromination.*)

Properties

The bromogermane prepared by this procedure has a molecular weight of 155.0 (calculated, 155.5) and a vapor pressure of 25.7 torr at $-23°$ (literature,[2] 25.0 torr). It is reported[2] to have a melting point of $-32°$ and a boiling point of $52°$. Its gas-phase infrared spectrum[12] shows characteristic bands at 2126, 2120 (r), 2110 (p), 871, 838 (r), 827 (p), and 578 cm.$^{-1}$. Its proton chemical shift[13] is $\tau5.50$ for a 5-10% v/v solution using cyclohexane as solvent and internal reference. Because bromogermane disproportionates slowly at room temperature into dibromogermane and germane, it is best stored in glass tubes kept at $-78°$.

Dibromogermane is a colorless liquid at room temperature with a molecular weight of 234.2 (calculated, 234.6), and a vapor pressure at $0°$ of 6.3 (literature,[2] 6.1 torr). Its melting and boiling points are $-15°$ and $89°$, respectively.[2] Sharp bands are observed at 2130 (p), 853 (r), 844 (p), 760 (r), and 751 cm.$^{-1}$ (p) in its gas-phase infrared spectrum.[14] In cyclohexane as solvent (5-10% v/v solution) and internal reference, dibromogermane has a proton chemical shift[13] of $\tau4.26$. It should be stored in a glass ampul cooled to $-78°$ because of its slow decomposition at room temperature.

References

1. C. H. Van Dyke, "Inorganic Derivatives of Germane and Digermane," in "Preparative Inorganic Reactions," W. L. Jolly (ed.), Vol. 6, Interscience Publishers, a division of John Wiley & Sons, Inc., New York, 1971.
2. L. M. Dennis and P. R. Judy, *J. Am. Chem. Soc.*, **51**, 2321 (1929).
3. K. M. Mackay, P. Robinson, E. J. Spanier, and A. G. MacDiarmid, *J. Inorg. Nucl. Chem.*, **28**, 1377 (1966).
4. S. Cradock and E. A. V. Ebsworth, *J. Chem. Soc.* (A), 1967, 12.
5. S. Cradock and E. A. V. Ebsworth, *ibid.*, **1967**, 1226.
6. T. N. Srivastava, J. E. Griffiths, and M. Onyszchuk, *Can. J. Chem.*, **40**, 739 (1962).
7. S. Sujishi and S. Witz, *J. Am. Chem. Soc.*, **16**, 4631 (1954).
8. W. L. Jolly and J. E. Drake, *Inorganic Syntheses*, **7**, 36 (1963).
9. D. F. Shriver, "The Manipulation of Air-sensitive Compounds," McGraw-Hill Book Company, New York, 1969.
10. W. L. Jolly, "The Synthesis and Characterization of Inorganic Compounds," p. 139, Prentice-Hall, Inc., Englewood Cliffs, N.J., 1970.

11. A. L. Beach and J. E. Griffiths, *Can. J. Chem.,* **44**, 743 (1966).
12. J. E. Griffiths, T. N. Srivastava, and M. Onyszchuk, *Can. J. Chem.,* **40**, 579 (1962).
13. E. A. V. Ebsworth, S. G. Frankiss, and A. G. Robiette, *J. Mol. Spectry.,* **12**, 299 (1964).
14. T. N. Srivastava, J. E. Griffiths, and M. Onyszchuk, *Can. J. Chem.,* **41**, 2101 (1963).

35. CHLOROGERMANE AND IODOGERMANE
(*Germyl Chloride and Germyl Iodide*)

Submitted by S. CRADOCK*
Checked by M. A. RING,† P. ESTACIO,† and M. D. SEFCIK†

Chlorogermane was first reported in 1929,[1] having been prepared from GeH_4 and HCl with an $AlCl_3$ catalyst. Subsequent methods have involved the reactions of BCl_3[2] or of hot $AgCl$[3] with GeH_4. All these methods result in the formation of GeH_2Cl_2 in addition to the desired product, which can be purified only by tedious fractionation. By contrast, the reaction of GeH_4 with $SnCl_4$ proceeds smoothly at room temperature[4] to yield GeH_3Cl uncontaminated by the dichloride. The reaction is slow, but acceptable yields are obtained by allowing it to proceed overnight.

Chlorogermane is a valuable starting material for the synthesis of other GeH_3 derivatives. Bromogermane and iodogermane may be prepared[5] cleanly and quickly by the reaction of GeH_3Cl with excess HBr or HI.

A. CHLOROGERMANE

$$GeH_4 + SnCl_4 \longrightarrow GeH_3Cl + HCl + SnCl_2$$

Procedure

The reaction takes place when germane[6] and tin(IV) chloride (tetrachlorostannane) are mixed at room temperature. It does not

---

*Department of Chemistry, University of Edinburgh, West Mains Road, Edinburgh, EH9 1HD, Scotland.
†Department of Chemistry, San Diego State University, San Diego, Calif. 92115.

appear to be accelerated markedly by moderate increase of temperature or by irradiation by visible light. Reaction appears to take place in a film of liquid $SnCl_4$ on the surface of the vessel. The $SnCl_2$ produced forms an adherent film. The reaction is retarded by the HCl produced, and the latter must be removed periodically if complete conversion of the germane is required. The presence of $SnCl_2$ formed in earlier reactions does not appear to affect the rate of reaction. No GeH_2Cl_2 is formed unless the reaction mixture is left for several days. The "initial" yield of GeH_3Cl depends strongly on the pressure of germane; overnight reactions produce initial yields of up to 50% with an initial germane pressure of 2 atmospheres but only 10–15% with 0.5 atmosphere. The over-all yields of GeH_3Cl may be enhanced by recycling the recovered germane after HCl has been removed by washing with water.

All manipulations may be carried out in a standard glass vacuum system using greased stopcocks; the reaction vessel itself is required to withstand internal pressure of 2–3 atmospheres and is best fitted with a greaseless stopcock. ■ **Caution.** *Shield well when under pressure.*

Pure germane* (GeH_4; 3.0 g., 39.2 mmoles) is condensed onto a wide cold finger in an evacuated 500-ml. reaction vessel, together with excess stannic chloride ($SnCl_4$; 6 ml. liquid, *ca.* 13.5 g., 52 mmoles), and the stopcock is closed. The cold finger is allowed to warm to room temperature, and the ampul is rotated to establish a film of liquid $SnCl_4$ over the surface. A film of white solid $SnCl_2$ begins to form within a few minutes. After 15 hours the reaction vessel is opened to the vacuum line† and the contents are fractionated using baths at $-64°$ (chloroform slush), $-120°$ (diethyl ether slush), and $-196°$. The first trap retains excess $SnCl_4$, the second pure GeH_3Cl (2.11 g., 19.0 mmoles; 48% initial yield). The third trap contains a mixture of GeH_4 and HCl. This mixture is condensed into an ampul containing distilled water and allowed to warm to room temperature. The ampul is shaken well and its most volatile contents fractionated at $-64°$. The fraction passing this trap is pure germane

---

*Available from Matheson Gas Products, East Rutherford, N.J. 07073.

†The checkers experienced a very low yield in one run in which the reaction vessel was cooled to $-196°$ before opening to the vacuum line. Chlorogermane may have been decomposed by condensation on the $SnCl_2$.

(infrared spectrum) and is returned to the reaction vessel with the excess $SnCl_4$.

If the reaction is scaled down, it is important that the volume of the reaction vessel be reduced proportionately to ensure a high initial conversion.

Properties

Chlorogermane is a mobile colorless liquid when condensed in a vacuum system (m.p. $- 52°$, b.p. 28°).[1] It is best characterized by its vapor pressure at 0°, 234 torr, and by its infrared spectrum[7] in the vapor phase. It is not spontaneously flammable or explosive in air and reacts only slowly with water, giving uncharacterized solid products. It should be stored at $- 196°$.

B. IODOGERMANE

$$GeH_3Cl + HI \longrightarrow GeH_3I + HCl$$

Procedure

Pure chlorogermane (GeH_3Cl; 0.145 g., 1.30 mmoles) and hydrogen iodide (0.186 g., 1.45 mmoles) are condensed together in a 100-ml. ampul with stopcock. (The checkers suggest that a mercury-free vacuum system be used for this preparation.) The ampul is allowed to warm to room temperature and is shaken to ensure mixing. The contents are fractionated using baths at $- 78°$ (acetone–solid CO_2) and $- 196°$. The $- 78°$ trap contains pure iodogermane (0.231 g., 1.14 mmoles), the other a mixture of HCl and HI.

A similar reaction using HBr in place of HI converts chlorogermane into bromogermane. Both reactions may be conveniently scaled up or down.

Properties

Iodogermane is very similar to the chloro derivative but is less volatile and has a higher melting point (m.p. $- 15°$; extrapolated b.p. *ca.* 90°; v.p. at 0°, 20 torr). It is rather less stable at room temperature than chlorogermane and is best handled in clean glassware with subdued lighting. It should be stored at $- 196°$.

References

1. L. M. Dennis and P. R. Judy, *J. Am. Chem. Soc.,* **51**, 2321 (1929).
2. See J. E. Drake, R. T. Hemmings, and C. Riddle, *J. Chem. Soc. (A),* **1970**, 3359.
3. K. M. Mackay, P. Robinson, E. J. Spanier, and A. G. MacDiarmid, *J. Inorg. Nucl. Chem.,* **28**, 1377 (1966).
4. J. E. Bentham, S. Cradock, and E. A. V. Ebsworth, *Inorg. Nucl. Chem. Letters,* **7**, 1077 (1971).
5. S. Cradock and E. A. V. Ebsworth, *J. Chem. Soc. (A),* **1967**, 12.
6. A. D. Norman, J. R. Webster, and W. L. Jolly, *Inorganic Syntheses,* **11**, 170 (1968).
7. D. E. Freeman, K. H. Rhee, and M. K. Wilson, *J. Chem. Phys.,* **39**, 2908 (1963).

36. FLUOROGERMANE AND DIGERMYLCARBODIIMIDE

$$2GeH_3Br + PbF_2 \longrightarrow 2GeH_3F + PbBr_2$$

$$2GeH_3F + (CH_3)_3SiNCNSi(CH_3)_3 \longrightarrow 2(CH_3)_3SiF + GeH_3NCNGeH_3$$

Submitted by S. CRADOCK*
Checked by J. BULKOWSKI,[†] G. NICKEL,[†] D. SEPELAK,[†] and C. VAN DYKE[†]

The interaction of trisilylamine with protonic acids is known to be a very useful reaction for preparing a variety of silyl derivatives.[1] Unfortunately, this method cannot be used in the synthesis of the analogous germyl derivatives, owing, in part, to the great thermal instability of $(GeH_3)_3N$.[2] An alternative germyl intermediate which reacts cleanly with protonic acids is digermylcarbodiimide, GeH_3-$NCNGeH_3$.[3] This reagent is reasonably stable thermally, and its interaction with H_2O, CH_3OH, and H_2S has been used in the synthesis of $(GeH_3)_2O$, GeH_3OCH_3, and $(GeH_3)_2S$, respectively.[3,4] In each case the only other product of the reaction is an involatile white solid, presumably polymerized cyanamide, $(H_2NCN)_x$.

Digermylcarbodiimide is prepared easily by the reaction of fluorogermane with bis(trimethylsilyl)carbodiimide. The general reaction of fluorogermane (or certain other halogermanes) with a

*Department of Chemistry, University of Edinburgh, West Mains Road, Edinburgh, EH9 1HD, Scotland.
†Department of Chemistry, Carnegie-Mellon University, Pittsburgh, Pa. 15213.

trimethylsilyl (or silyl) derivative exemplifies an important synthetic procedure for the preparation of various other germyl derivatives.[5]

The preparation of GeH_3F itself presents some difficulties. It is by far the least stable of the halogermanes. It decomposes to form GeH_4 and GeH_2F_2 initially and forms polymeric products at room temperature. As a result, the fluorination of other halogermanes must be carried out in the mildest possible manner. The original preparation of GeH_3F[6] involved the use of AgF, which is a vigorous fluorinating agent and is difficult to prepare and store in a dry state. It has been found that freshly prepared lead(II) fluoride fluorinates bromogermane quite cleanly.[7]

A. FLUOROGERMANE

Materials

The success of converting GeH_3Br to GeH_3F depends on the state of the PbF_2 used. It must be freshly precipitated and dried to a free-flowing powder without strong heating. The precipitate formed by mixing equivalent quantities of $Pb(NO_3)_2$ and NaF in aqueous solutions is filtered and washed well with distilled water and then with acetone. It is allowed to dry in air at a temperature below 100° until all lumps can be crushed and then overnight at this temperature. The resulting white "tacky" powder is dried under a high vacuum until it is free-flowing and no further water is evolved. Finally it is ground in a mortar with an equal volume of powdered glass.

The synthesis of GeH_3Br is given on p. 157 of this volume.

Procedure

A mixture of *ca.* 85 g. of lead(II) fluoride with glass powder is packed into a reaction tube (15 mm. o.d. × 200 mm. long)* fitted with standard-taper joints and with the ends plugged with glass wool.

---

*The checkers used a glass column that had a bypass tube (with a stopcock) leading from the top of the column to the bottom sample chamber. The GeH_3Br could then be distilled directly from the vacuum line to the bottom of the column through the bypass. The conversion was accomplished by closing the bypass stopcock and allowing the GeH_3Br to pass through the packed column into a − 196° trap via a − 78° trap.

It is essential that the packing be loose enough to allow gas flow through the tube but not loose enough to allow settling of the contents and formation of open channels. The tube is attached to a vacuum system at one end, and an ampul with stopcock containing 2.16 g. (13.9 mmoles) of bromogermane is attached to the other end. The tube is evacuated thoroughly with the GeH_3Br condensed at $-196°$. The stopcock is opened, and the bromogermane is allowed to evaporate through the tube into the vacuum system, where the products are passed through traps at $-78°$ and $-96°$ into one at $-196°$.

The $-78°$ trap contains GeH_2F_2, some unreacted bromogermane, and any water that may be evolved. The $-96°$ trap retains additional bromogermane (identified by infrared peaks at 838 and 827 cm.$^{-1}$). It is advantageous to use detachable traps so that the bromogermane-containing mixtures need not be pumped to waste. Instead they may be recycled through lead(II) fluoride for further conversion to fluorogermane.

The $-196°$ trap contains GeH_3F, GeH_4, and a trace of SiF_4; a single fractionation at $-120°$ (diethyl ether slush bath) retains pure GeH_3F and allows GeH_4 and SiF_4 to pass. The purity of the fluorogermane is established using its infrared spectrum[8,9] and vapor density.[6] The yield of fluorogermane is 11.8 mmoles or 83%. Germane (GeH_4; 1.2 mmoles) is a by-product.

Properties

Fluorogermane is distinctly unstable at room temperature and must be stored at low temperature, preferably at $-196°$ (liquid nitrogen). The extrapolated boiling point of the compound is 15.6°; its melting point is $-22°$.[6] Calculated vapor pressures for GeH_3F are 359 torr at 0° and 100 torr at $-22.8°$ (CCl_4 slush).[6] Fluorogermane is more susceptible to hydrolysis than the other germyl halides but is not spontaneously inflammable in air.

B. DIGERMYLCARBODIIMIDE

$$CaCN_2 + 2AgNO_3 \longrightarrow Ag_2CN_2 + Ca(NO_3)_2$$

$$2(CH_3)_3SiCl + Ag_2CN_2 \longrightarrow (CH_3)_3SiNCNSi(CH_3)_3 + 2AgCl$$

$$(CH_3)_3SiNCNSi(CH_3)_3 + 2GeH_3F \longrightarrow GeH_3NCNGeH_3 + 2(CH_3)_3SiF$$

Materials

■ **Caution.** *Silver cyanamide is a potentially explosive material.* Silver cyanamide is prepared by the following procedure. Calcium cyanamide (Eastman, 100 g.) is added in small portions to 1 l. of 8% w/w H_2SO_4 (44 ml. of concentrated H_2SO_4 diluted to 1 l. with distilled water). The solution is kept below 40° during the addition by cooling with ice. After filtration, the clear yellow solution is treated with 200 g. of silver nitrate in 500 ml. of distilled water. Any yellow silver cyanamide precipitate formed at this stage is filtered. The solution is neutralized with aqueous ammonia, and the yellow-orange precipitate is recovered by filtration. The combined precipitates are washed with water, then with methanol, and are dried *in vacuo without heating* (yield *ca.* 70 g.).

Bis(trimethylsilyl)carbodiimide is prepared by a modification of the method of Pump and Wannagat[10] applicable to work under vacuum-line conditions. Chlorotrimethylsilane [$(CH_3)_3SiCl$; 2.9 g., 26.8 mmoles] is condensed onto thoroughly dried silver cyanamide (Ag_2CN_2; 3.4 g., 13.3 mmoles) in an evacuated ampul, which is allowed to warm slowly to room temperature and left overnight.

■ **Caution.** *While no accidents have ever been encountered at this point, it is considered advisable, in view of the potentially explosive nature of the silver salt Ag_2CN_2 to take full precautions against explosions.* The ampul is then opened to the vacuum line and the volatile contents fractionated using traps at $-22°$ (CCl_4 slush) and $-196°$. The first trap contains the desired product contaminated with hexamethyldisiloxane, which is removed by repeatedly passing the product into a trap at $-22°$. The disiloxane passes through to a liquid-nitrogen trap. The purity of the product is established by the absence of the strong infrared band of hexamethyldisiloxane at 1067 cm.$^{-1}$ in its vapor-phase infrared spectrum.

Chlorotrimethylsilane is available from a number of commercial sources (e.g., Columbia Organic Chemicals, Columbia, S.C. 29205).

Procedure

Fluorogermane (GeH_3F; 0.570 g., 6.03 mmoles) is condensed into an ampul with stopcock with bis(trimethylsilyl)carbodiimide [$(CH_3)_3SiNCNSi(CH_3)_3$; 0.554 g., 2.98 mmoles]. The ampul is closed and allowed to warm to room temperature. The ampul is then shaken for a few seconds to ensure mixing and is refrozen in liquid nitrogen. Fractionation of the products using traps at $-122°$ and $-196°$ gives the product as a colorless solid in the first trap, while $(CH_3)_3SiF$ and excess GeH_3F collect in the second trap. The product, $GeH_3NCNGeH_3$ (0.503 g., 2.63 mmoles, 88% yield), may be finally purified by a second transfer into a trap at $-22°$. Its purity may be judged from its vapor-phase infrared spectrum,[8,9] its vapor pressure (~ 0.6 torr at $0°$), and its melting point ($10.0 \pm 0.5°$).

Properties

Digermylcarbodiimide is a rather involatile liquid, having a vapor pressure of about 2 torr at $20°$.[3] The compound decomposes slowly above $0°$ and must be stored at low temperatures, preferably at $-196°$ (liquid nitrogen). It may be handled by normal vacuum-line techniques in clean glassware but is decomposed readily by water and protonic acids.

References

1. E. A. V. Ebsworth and J. C. Thompson, *J. Chem. Soc. (A),* **1967**, 69.
2. D. W. H. Rankin, *J. Chem. Soc. (A),* **1969**, 1926.
3. S. Cradock and E. A. V. Ebsworth, *J. Chem. Soc. (A),* **1968**, 1423.
4. S. Cradock, E. A. V. Ebsworth, and D. W. H. Rankin, *J. Chem. Soc. (A),* **1969**, 1628.
5. C. H. Van Dyke, "Inorganic Derivatives of Germane and Digermane," in "Preparative Inorganic Reactions," W. L. Jolly (ed.), Vol. 6, p. 157, Interscience Publishers, a division of John Wiley & Sons, Inc., New York, 1971.
6. T. N. Srivastava and M. Onyszchuk, *Proc. Chem. Soc.,* **1961**, 205.
7. E. A. V. Ebsworth, S. G. Frankiss, and A. G. Robiette, *J. Mol. Spectry.,* **12**, 294 (1964).
8. D. E. Freeman, K. H. Rhee, and M. K. Wilson, *J. Chem. Phys.,* **39**, 2908 (1963).
9. J. E. Griffiths, T. N. Srivastava, and M. Onyszchuk, *Can. J. Chem.,* **40**, 579 (1962).
10. J. Pump and U. Wannagat, *Annalen,* **652**, 421 (1962).

37. IODODIGERMANE AND METHYLDIGERMANE

$$Ge_2H_6 + I_2 \longrightarrow Ge_2H_5I \xrightarrow{CH_3MgI} Ge_2H_5CH_3$$

Submitted by KENNETH M. MACKAY*
Checked by J. E. DRAKE† and R. T. HEMMINGS†

The lower germanes, GeH_4, Ge_2H_6, and Ge_3H_8, are readily prepared[1,2] and well suited for study of the hydride series. Although their thermal stabilities are somewhat lower than those of the polysilanes, this is offset by their less violent reaction with air. Routes for the formation of substituted germanes are well established, but introduction of a substituent into digermane or trigermane has been achieved by only a limited number of reactions.[3]

Reaction of digermane with iodine gives Ge_2H_5I in high yield[4] if the reaction is carried out at low temperatures in absence of a solvent, as illustrated by the recovery[5] of Ge_2H_5D in better than 90% yield after treatment *in situ* with $Li[AlD_4]$. However, iododigermane is thermally unstable and relatively involatile, and on distillation in a vacuum line at $0°$ losses of about 20% are to be expected. Even in the vapor phase at 4 mm. pressure, Ge_2H_5I is almost completely decomposed after 2 hours at room temperature.

Thus the preferred procedure is to use the iodide without isolating it. As an example, we give the preparation of methyldigermane. Ethyldigermane[6] and pentacarbonyl(digermanyl)manganese[7] have been prepared similarly. In contrast to the halides, all these species have reasonable thermal stability. If a digermanyl halide is desired, the more volatile and somewhat more stable chloride is preferable.

Similar remarks apply to trigermane where the iodide cannot be isolated but may be converted[8] to Ge_3H_7D, or to methyltrigermane, in high yield. If the iodine reaction is carried out at $-63°$ or at $-45°$, substitution[9] takes place primarily on the central germanium.

---

*School of Science, University of Waikato, Hamilton, New Zealand.
†Department of Chemistry, University of Windsor, Windsor 11, Ontario, Canada.

A. IODODIGERMANE

$$Ge_2H_6 + I_2 \longrightarrow GeH_3GeH_2I + HI$$

Procedure

■ **Caution.** *Germanes are generally regarded as toxic. Although the higher hydrides do not usually inflame spontaneously in air, they are readily ignited at temperatures as low as 50°, and combustion is rapid or explosive.*

Preparations may be carried out in a conventional vacuum line. Germanes react slowly with mercury, and manometers should be isolated when not in use. Hydrocarbon stopcock lubricants dissolve germanes, but decomposition is slow. Glassware coming in contact with halogermanes should be new or carefully cleaned from solid residues. A penultimate acid-washing stage is advisable in cleaning. The reaction is carried out conveniently in a 2-cm.-diam. tube attached to the vacuum line via a 24-mm. socket and equipped with a 14-mm. socket pointing up at a 45° angle and at least 15 cm. above the bottom. The most convenient scale of working is 1 to 2 mmoles; scaling up may lead to considerably reduced yields.

Resublimed iodine (242 mg., 0.95 mmole) is placed in the reaction tube and flushed with dry nitrogen. With the side arm stoppered, the tube is attached to the line and evacuated, with cooling to prevent loss of iodine. Digermane[1] (151 mg., 1.0 mmole, 5% excess) is condensed directly onto the iodine using liquid nitrogen, and then the mixture is warmed to − 63.5° (chloroform slush bath) and held at that temperature. [The checkers found that a − 78° bath (Dry Ice–acetone) was more convenient.] Complete reaction, shown by the disappearance of the iodine, takes up to 9 hours at − 63° and leaves a pale-green reaction mixture. Alternatively, the reaction may be accelerated by occasional slight local warming with the fingers. In this manner, the reaction is complete in 1–2 hours at the price of slightly lower yield of mono-substituted product and usually leaves a dark-brown reaction mixture. When the iodine has disappeared, by-products volatile at − 63.5° are distilled, leaving a pale-yellow product. Completeness of reaction is conveniently checked by weighing the total − 63.5° volatile fraction (calcd. 126 mg. = 118.6

mg. of HI plus 7.6 mg. of excess Ge_2H_6). It is normal to find more than 95% of the calculated weight. Alternatively, the HI may be fractionated at $-116°$.

At this stage, there remains in the reaction vessel Ge_2H_5I contaminated by traces of germyl iodide and small amounts of polyiododigermanes, mostly $IGeH_2GeH_2I$. This sample is suitable for conversion to other digermanyl compounds or may be purified by pumping at $-22°$ (carbon tetrachloride slush) to remove GeH_3I and then distilled from a 0° bath under as good a vacuum as possible. Provided the sample is in the liquid phase for the shortest possible time, the yield may reach 90%. However, 50–60% is more usual, and the yield may drop much lower. The loss occurs almost entirely at the distillation step. (■ **Note.** *The iododigermane is quite unstable thermally and should be stored at liquid-nitrogen temperature under a layer of frozen dibutyl ether until it is to be used.*)

Properties

Iododigermane is colorless and nonvolatile at $-22°$ and melts at about $-17°$. The liquid rapidly turns yellow with evolution of germane, some digermane, and hydrogen. It may be distilled under vacuum at 0° with 70–90% recovery. Addition[4] of dry, powdered AgBr or AgCl followed by distillation at 0° gives Ge_2H_5Br (5–15% yield) or Ge_2H_5Cl (18–61% yield), respectively. The halides are readily characterized[5] by the pair of sharp, strong deformation modes at 680–800 cm.$^{-1}$ in the infrared spectrum. The strong band at 785–795 cm.$^{-1}$ is assigned as the GeH_3 symmetric deformation and is almost constant in position for the three halides, while the very strong mode described as the GeH_2 wag occurs at 721 cm.$^{-1}$ (Ge_2H_5Cl), 705 cm.$^{-1}$ (Ge_2H_5Br), or 680 cm.$^{-1}$ (Ge_2H_5I). The first sign of decomposition is the appearance of the very strong digermane doublet at 760 cm.$^{-1}$.

Chlorodigermane. If it is essential to manipulate a digermanyl halide, it is preferable to use the more stable chloride. In addition to the synthesis from the iodide, the chloride may be made from the reaction of digermane with heated AgCl[4] or with a deficit of $SnCl_4$ at room temperature according to the equation:[11]

$$Ge_2H_6 + SnCl_4 \longrightarrow Ge_2H_5Cl + HCl + SnCl_2$$

The yield from any of these preparative routes is highly variable but may reach 70%.

Ge_2H_5Cl has a vapor pressure of 0.7 mm. at $-45.2°$ and 1.8 mm. at $-21.9°$ (CCl_4 slush). It may be distilled at these temperatures with only slight loss. At room temperature, digermane may be detected after 30 minutes, and a gas-phase sample at 4 mm. pressure completely decomposes in 6 to 9 hours. Chlorodigermane should be stored in a sealed glass tube kept in liquid nitrogen or in solid CO_2.

B. METHYLDIGERMANE

$$GeH_3GeH_2I + CH_3MgI \longrightarrow GeH_3GeH_2CH_3 + MgI_2$$

Procedure

Methylmagnesium iodide (iodomethylmagnesium) solution (p. 185 or Vol. IX, p. 92) is prepared under nitrogen in di-*n*-butyl ether. ■ **Caution.** *Di-n-butyl ether is very prone to form explosive peroxides. It should be treated first with activated alumina, then with sodium, distilled (b.p. 142°), and stored under nitrogen.* Care should be taken to consume all the iodomethane used in the reaction, because it is difficult to separate from methyldigermane. A twofold excess of the Grignard reagent (relative to the Ge_2H_5I to be used) is transferred under nitrogen into a suitable angled tube fitted with an extended 14-mm. cone.

Dry nitrogen is passed through the vacuum line into the reaction vessel containing the frozen iodide. When the nitrogen pressure reaches atmospheric, the stopper is removed from the side arm and a steady flow of nitrogen is continued until the tube containing the CH_3MgI solution has been put in place. The nitrogen flow is stopped, the Grignard solution is frozen, and the whole system is evacuated. The Grignard solution then is allowed to melt and is outgassed and added to the iodide by rotating the side tube about the 14-mm. joint. The liquid nitrogen around the reaction vessel is replaced by a $-45.2°$ bath (chlorobenzene slush) and the reagent is allowed to warm to $-45°$. Two layers will form which react steadily at the interface. After 15 minutes at $-45°$, the bath is removed and the mixture is allowed to warm to room temperature to complete the reaction, shown by the disappearance of the iodide layer. On freezing in liquid nitrogen, no more than a trace of noncondensable gas should be observed. The reaction mixture is then allowed to warm, and about a fifth of the volume is distilled out through traps held at $-63.5°$, $-116°$ (diethyl ether slush or the safer n-propanol slush melting about $-120°$ may be substituted), and $-196°$.* Traces only of ethane and methylgermane should be present in the liquid-nitrogen trap. The dibutyl ether is held at $-63.5°$, and the head fraction from this trap may show the presence of dimethyldigermane. Refractionation of the $-116°$ fraction through two traps at $-78°$ (solid CO_2) removes dibutyl ether to give pure methyldigermane in 60–80 % yield over-all, based on the iodine used in the initial reaction.

An alternative and more convenient separation is to distill about 20% of the reaction mixture into a septum tube and separate[10] by g.l.c. using a silicone active phase. This procedure avoids distillation losses, allows ready separation from methyl iodide, and allows the recovery of polymethyldigermanes.

By following the same procedure, but allowing the iodine to react with digermane at higher temperatures,[10] enhanced yields of poly-methyl derivatives result. For example, Ge_2H_6 (1.26 mmoles) plus I_2 (1.21 mmoles) held at $-45°$ for 2 hours and then treated with CH_3MgI (4 mmoles) gave $CH_3GeH_2GeH_3$ (66 %), $(CH_3)_2GeHGeH_3$ 1 %), $CH_3GeH_2GeH_2CH_3$ (12 %), and $(CH_3)_2GeHGeH_2CH_3$ (1.5 %). For such an experiment, g.l.c. separation is essential. Ethyldigermanes[6] or alkyltrigermanes[9] are also prepared by this procedure.

Properties

Methyldigermane is a colorless, relatively stable liquid[6] whose vapor pressure obeys the equation

$$\log p \text{ (mm.)} = \frac{-1637.7}{T} + 7.877$$

over the range of -63 to $+24°$. The boiling point (extrapolated) is $54.7°$, and Trouton's constant $= 22.9$ cal. deg.$^{-1}$ mole^{-1}. The vapor pressure at $-63.5°$ was 0.99 mm. and at $0.1°$ was 74.1 mm.

Like other germanes, methyldigermane dissolves in stopcock grease and reacts slowly with mercury. Ideally, it is handled in an all-glass or glass-Teflon system, but little difficulty will be experienced in using a conventional vacuum line for any of the alkyldigermanes.

---

*In trap-to-trap distillation, the best temperatures for separation can vary according to such factors as efficiency of the vacuum, volume of liquid, and vacuum-line construction.

References

1. W. L. Jolly and J. E. Drake, *Inorganic Syntheses*, 7, 34 (1963).
2. A. D. Norman, J. R. Webster, and W. L. Jolly, *Inorganic Syntheses*, 11, 176 (1968).
3. C. H. Van Dyke, in "Preparative Inorganic Reactions," W. L. Jolly (ed.), Vol. 6, p. 157, Interscience Publishers, a division of John Wiley & Sons, Inc., New York, 1971.
4. K. M. Mackay, P. Robinson, E. J. Spanier, and A. G. MacDiarmid, *J. Inorg. Nucl. Chem.*, 28, 1377 (1966).
5. K. M. Mackay, P. Robinson, and R. D. George, *Inorg. Chim. Acta*, 1, 236 (1967).
6. K. M. Mackay, R. D. George, P. Robinson, and R. Watt, *J. Chem. Soc. (A)*, 1968, 1920.
7. S. R. Stobart, *Chem. Commun.*, 1970, 999.
8. K. M. Mackay and P. Robinson, *J. Chem. Soc.*, 1965, 5121.
9. S. T. Hosfield and K. M. Mackay, *J. Organometallic Chem.*, 24, 107 (1970).
10. R. D. George and K. M. Mackay, *J. Chem. Soc. (A)*, 1969, 2122.
11. K. M. Mackay and R. Watt, unpublished observations, 1966.

38. PENTACARBONYLGERMYLMANGANESE

$$GeH_3Br + Na[Mn(CO)_5] \longrightarrow H_3GeMn(CO)_5 + NaBr$$

Submitted by STEPHEN R. STOBART*
Checked by E. A. V. EBSWORTH† and A. ROBERTSON†

Reactions of organometallic halides with anionic complexes of the metal carbonyls have been widely used to obtain metal-metal bonded compounds.[1] This general synthetic route has recently been successfully applied to the preparation of germyl-metal carbonyls. It should be possible to extend the present range of such complexes, those of manganese,[2,2a] rhenium,[3] iron,[4] and cobalt,[5] to other transition metals by using appropriate carbonyl anions. Bromogermane[6] has been used as a convenient source of the GeH_3 group. Its reaction with sodium pentacarbonylmanganate, to give pentacarbonylgermylmanganese, $GeH_3Mn(CO)_5$, is described here and is illustrative of this type of preparation, giving in high yield a

*Department of Chemistry, Queen's University, Belfast, BT9 5AG, Northern Ireland.
†Department of Chemistry, University of Edinburgh, Edinburgh EH9 3JJ, Scotland.

product which can be purified readily using standard vacuum techniques.

Procedure

■ **Caution.** *Attention is drawn to the toxicity of metal carbonyls. Volatile metal carbonyl derivatives are likely to be particularly dangerous and should be handled only in a vacuum system or using an efficient fume hood. Hazards arising during the purification of tetrahydrofuran, required as a solvent for this reaction, have been emphasized in a previous volume of this series (Vol. XII, p. 317).*

The synthesis of $Na[Mn(CO)_5]$ has been described in detail by King and Stone.[7] Sodium amalgam formed by addition of 0.29 g. (12.5 mmoles) of sodium to 4 ml. (54 g.) of mercury* is reacted with 1.800 g. (4.62 mmoles) of decacarbonyldimanganese,† $Mn_2(CO)_{10}$, in 30 ml. of dry tetrahydrofuran‡ under a slow flow of dry nitrogen. Vigorous stirring is necessary to ensure frequent breaking of the surface of the amalgam. The reaction takes approximately 2 hours, after which time the solution is greenish. At this stage excess sodium is removed by "washing" with 3 ml. of clean mercury as described previously.[7] The reaction vessel containing the solution of the anion is fitted with an appropriate adapter, attached to a vacuum system, and two-thirds of the tetrahydrofuran is removed by pumping. (If possible, grease-free stopcocks should be used in the vacuum system, because the product is quite soluble in conventional greases.)

Simultaneously with the above, the reaction between germane and bromine (described elsewhere in this section, p. 157) is used to prepare 1.000 g. (6.43 mmoles) of bromogermane, purified on a vacuum line by fractionation through a trap at $-46°$ (chlorobenzene slush) and condensation at $-95°$ (toluene slush).

The reaction is conducted as follows: the flask containing the manganate solution is cooled to $-196°$ (liquid nitrogen) and

---

*The sodium should be added in small pieces, allowing each piece to react (heat evolution!) before the next is added. It may be necessary to crush the first piece under mercury in order to initiate reaction.

†Available from Strem Chemicals, Inc., Danvers, Mass. 01923.

‡Alternatively, dry diethyl ether may be used; however, its higher volatility, advantageous in subsequent fractionation operations, is offset by a lowering of solubility for most metal carbonyls and carbonyl anions.

evacuated thoroughly, after which the bromogermane is distilled into the solution under vacuum. The mixture is warmed to room temperature during 15 minutes and then is stirred gently magnetically for a further 30 minutes, during which time the green color lightens and a white solid is deposited. Next, volatile components are pumped out of the reaction flask for about 2 hours, through a fractionation line with traps cooled to $-25°$ (*o*-xylene slush), $-46°$, $-78°$, and $-196°$. (The solid CO_2-acetone trap at $-78°$ condenses tetrahydrofuran without solidifying it and thus blocking the vacuum line; traces of germane are recovered at $-196°$.) Combination of the $-25°$ and $-46°$ fractions is followed by refractionation through a trap held at $-46°$. This procedure affords a condensate* which is spectroscopically pure pentacarbonylgermylmanganese (1.505 g., 5.55 mmoles) as a creamy-white solid, m.p. 24°, identifiable by its gas-phase infrared and by its ^1H n.m.r. spectra. The yield based on bromogermane is 86%.

Properties

Pentacarbonylgermylmanganese is a white, low-melting solid, and is sufficiently volatile for easy vacuum manipulation. Vapor pressures in the range 32–81° determine the equation log p(mm.) $= -2300/T + 8.3$. The infrared spectrum of the vapor contains[2] only the following bands above 600 cm.$^{-1}$: 2115 (s); 2063 (s) (sh); 2022, 2019 (vvs); 1981 (m) [due to v(CO) and v(GeH)]; 821, 818, 815 (s) (RQP) [due to δ(GeH$_3$)]; and 663 (vs) [due to δ(MnCO)]. The ^1H n.m.r. spectrum consists[2] of a sharp singlet in benzene solution, at 3.28 p.p.m. down field from tetramethylsilane.

References

1. R. E. J. Bichler, M. R. Booth, H. C. Clark, and B. K. Hunter, *Inorganic Syntheses*, **12**, 60 (1970).
2. K. M. Mackay and R. D. George, *Inorg. Nucl. Chem. Letters*, **5**, 797 (1969).
2a. S. R. Stobart, *Chem. Commun.*, **1970**, 999.
3. K. M. Mackay and S. R. Stobart, *Inorg. Nucl. Chem. Letters*, **6**, 687 (1970).

*Unless the initial distillation is conducted very slowly, traces of GeH$_3$Mn(CO)$_5$ are carried through by tetrahydrofuran; if desired, this additional material may be recovered by repeated fractionation of the contents of the $-78°$ trap.

4. S. R. Stobart, *ibid.,* **7**, 219 (1971).
5. K. M. Mackay and R. D. George, *ibid.,* **6**, 289 (1970).
6. M. F. Swiniarski and M. Onyszchuk, *Inorganic Syntheses,* **15**, 157 (1974).
7. R. B. King and F. G. A. Stone, *ibid.,* **7**, 198 (1963).

39. GERMYLPHOSPHINE

$$LiAlH_4 + 4PH_3 \longrightarrow 4H_2 + Li[Al(PH_2)_4]$$

$$Li[Al(PH_2)_4] + 4GeH_3Br \longrightarrow LiBr + AlBr_3 + 4GeH_3PH_2$$

Submitted by A. D. NORMAN,* D. C. WINGELETH,* and C. A. HEIL*
Checked by C. RIDDLE† and W. L. JOLLY†

Germylphosphine can be prepared from the ozonizer discharge decomposition of germane-phosphine mixtures,[1] the acid hydrolysis of germanide-phosphide alloys,[2,3] the exchange reaction of bromogermane and silylphosphine,[4,5] and the phosphination of halogermanes[6,7] with lithium tetraphosphinoaluminate, $Li[Al(PH_2)_4]$. The first two methods are unsuitable for general synthesis because they result in relatively low yields of GeH_3PH_2. In addition, the complex product mixtures which form are tedious to separate. The SiH_3PH_2-GeH_3Br exchange reaction requires the initial synthesis of SiH_3PH_2—a compound which is as difficult to prepare as GeH_3PH_2. Thus, of the four available methods, the phosphination of germyl halides with $Li[Al(PH_2)_4]$ is clearly superior because it results in high product yields, is relatively safe, and yields the product in a mixture from which it is easily separated. Using this method, up to 20 mmoles of GeH_3PH_2 can be prepared conveniently in one reaction.

Procedure

■ **Caution.** *Germylphosphine is flammable and probably highly toxic. Therefore, it should be handled in a vacuum line or oxygen-free system at all times.*

*Department of Chemistry, University of Colorado, Boulder, Colo. 80302.
†Inorganic Materials Research Division, Lawrence Berkeley Laboratory, University of California, Berkeley, Calif. 94720.

1. Materials

Pure, white, crystalline lithium tetrahydroaluminate ($LiAlH_4$) is preferred for this reaction. The usual commercial material, a gray powder which is not completely soluble in ether, may be purified by extraction with diethyl ether.[8] Peroxide-free diethyl ether is used to extract the gray $LiAlH_4$ powder *in an inert atmosphere.* A Soxhlet extractor is convenient. The clear extract may then be evaporated under reduced pressure (no heating) to give a white, crystalline powder.

Diglyme [bis(2-methoxyethyl) ether] (b.p. 62–63° at 15 torr or 160° at atmospheric pressure) should be distilled from CaH_2[9] or Vitride or Red-Al reducing agents (p. 111). In no instance should the mixture be allowed to distill to dryness.

Phosphine is commercially available from the Matheson Co. or may be prepared by methods described in *Inorganic Syntheses.*[10,11] The commercial PH_3 should be purified by passing it through a − 130° (*n*-pentane slush) trap into a − 196° trap prior to use.

Bromogermane is prepared as described on p. 157.

2. Apparatus

The synthesis is carried out in a 500-ml., single-necked, round-bottomed flask which is connected to a standard vacuum line by means of a stopcock adapter. The stopcock adapter is important, since it allows for the easy attachment and removal of the reaction vessel from the vacuum line without exposing the contents of the reactor to air. The vacuum line must be equipped with a mercury manometer in order to measure the pressure of PH_3 during the preparation of the $Li[Al(PH_2)_4]$ solution. A conventional vacuum line with greased stopcocks is satisfactory. However, germyl-phosphine does attack conventional stopcock lubricants; so a greaseless system is preferred if available.

3. Preparation of $Li[Al(PH_2)_4]$ Solutions

The reaction vessel is charged with 0.19 g. (5.0 mmoles) of pure $LiAlH_4$, 10 ml. of freshly distilled diglyme, and a glass-covered magnetic stirring bar. By means of the stopcock adapter the reaction flask is attached to the vacuum line. This should be done as quickly as possible in order to minimize exposure to air. All joints should be greased with Dow-Corning high-vacuum silicone grease in order to eliminate the slow dissolution of the grease by diglyme vapor during the reaction. The reactor is cooled to 0°, and the bulk of the air is removed. The flask is cooled to $-196°$ (liquid nitrogen) with continual pumping. Repetition of the warming-cooling cycle while pumping on the reaction system results in the complete removal of air. The reaction vessel is maintained at 0°, and phosphine is allowed to diffuse into the reactor until a pressure of 500–550 torr is reached. The stopcock of the adapter is closed, the reaction vessel is allowed to warm to room temperature, and the reaction contents are stirred. Within a few minutes, reaction commences, as evidenced by the evolution of H_2 from the reaction mixture and the slow dissolution of undissolved $LiAlH_4$. It is recommended that rubber bands or springs be used to hold the greased joints firmly together during the preparation of the $Li[Al(PH_2)_4]$ solution. (Vacuum alone cannot be relied on to hold the joints together, owing to the buildup of hydrogen pressure.) After 6–8 hours the entire reaction vessel is cooled to $-196°$ and H_2 is pumped from the flask through a trap cooled to $-196°$ (to collect traces of PH_3). The flask is warmed slowly to 0°, and PH_3 is added again until a pressure of *ca.* 500 torr is reached in the system. The reaction vessel is allowed to warm to room temperature as described above. After 12 hours, H_2 is again removed at $-196°$. Repetition of this scheme of PH_3 addition and H_2 removal is continued until only 5–10 torr of H_2 pressure is observed after a 24-hour reaction interval. Typically this takes about 3 days, or 4 additions of PH_3. At the end of the reaction period, the solution of $Li[Al(PH_2)_4]$ should be colorless and free of solids.

4. Reaction of GeH₃Br with Li[Al(PH₂)₄] Solution

The $Li[Al(PH_2)_4]$ solution prepared as described above is freed of PH_3 by evacuation. Bromogermane (15 mmoles) is condensed at $-196°$ into the flask. The stopcock is closed, and the flask is warmed slowly. At about $-50°$ the diglyme solution begins to melt. The reaction vessel is maintained at $-45°$ (chlorobenzene slush) and allowed to stir for a period of 20 minutes. Upon completion of the reaction, the reaction vessel is warmed to 0° and the contents are pumped into the vacuum line through a $-30°$ trap (bromobenzene slush bath) into a $-196°$ trap. The GeH_3PH_2, PH_3 (generated in a side reaction), and GeH_4 (generated as a result of residual Al—H material in the $Li[Al(PH_2)_4]$ solution and slight decomposition of GeH_3PH_2) are condensed in the $-196°$ trap. Diglyme is condensed in the $-30°$ trap.

Pure GeH_3PH_2 is obtained by repeated (twice) passage of the $-196°$ condensate through a series of traps at $-78°$ (Dry Ice–acetone slush), $-130°$ (pentane slush), and $-196°$. The GeH_3PH_2 collects in the $-130°$ trap. A yield of about 10 mmoles (67%) can be expected. (The checkers obtained a 90% yield based on GeH_3Br by using an excess of $Li[Al(PH_2)_4]$.)

The residual material in the reaction vessel should be disposed of carefully. After allowing the mixture to stand at room temperature for 6–8 hours, the last traces of high-volatility materials are pumped from the reaction vessel. The flask can be removed from the vacuum line and cooled to $-196°$ and the adapter removed. This step should be performed in a good hood. After 20 ml. of isopropyl alcohol is added to the frozen reaction materials, and as they warm to room temperature, deactivation takes place. After 1–2 hours the deactivated mixture can be flushed down the hood sink. ■ **Caution.** *Since H_2 and small amounts of PH_3 will be evolved during this step, care should be taken to keep the reaction vessel in a good hood. Phosphine is highly toxic.*

This synthesis can be scaled up conveniently by a factor of 2 provided that a correspondingly larger reaction vessel is used. Also, the $Li[Al(PH_2)_4]$-phosphination method may be used for the preparation of phosphinogermanes in general. Details of the preparation of $(CH_3)_3GePH_2$, $[(CH_3)_2GeH]PH_2$, and $(CH_3)_2Ge$-$(PH_2)_2$ can be found in the literature.[12]

Properties

Germylphosphine has an exceptionally strong odor (similar to garlic) and is undoubtedly very poisonous. The vapor pressure may be represented by the equation $\log P_{mm.} = 7.277 - 1415.5/T$ (where T = kelvin). The extrapolated boiling point is $48.8°$. The $-45°$ vapor pressure (chlorobenzene slush) is 12.0 torr and may be used as a criterion of purity.[7] The infrared spectrum shows absorption at 2310 (s), 2093 (vs), 2086 (vs), 1073 (w), 888 (m), 815 (vs), and 701 (m). The mass spectrum shows peaks from m/e 31–34, 70–79, and 101–112, attributable to PH_x^+, GeH_x^+, and $GePH_x^+$ ions, respectively.

Germylphosphine decomposes slowly at room temperature in the gas phase. Therefore, it should be stored at reduced temperature (i.e., $-78°$ or $-196°$). Thermal decomposition yields PH_3, $(GeH_3)_3P$, and polymer. Germylphosphine can be separated readily from these products by passing it twice through a series of traps at $-78°$, $-130°$, and $-196°$. The GeH_3PH_2 is condensed in the $-130°$ trap.

References

1. J. E. Drake and W. L. Jolly, *Chem. Ind.,* **1962,** 1470.
2. P. Royen, G. Rocktaschel, and W. Mosch, *Angew. Chem. Int. Ed.,* **3,** 703 (1964).
3. P. Royen and C. Rocktaschel, *Z. Anorg. Allgem. Chem.,* **346,** 290 (1966).
4. J. E. Drake, N. Goddard, and J. Simpson, *Inorg. Nucl. Chem. Letters,* **4,** 361 (1968).
5. J. E. Drake, N. Goddard, and C. Riddle, *J. Chem. Soc. (A),* **1969,** 2704.
6. D. C. Wingeleth and A. D. Norman, *Chem. Commun.,* **1967,** 1218.
7. D. C. Wingeleth and A. D. Norman, *Inorg. Chem.,* **9,** 98 (1970).
8. L. F. Fieser and M. Fieser, "Reagents for Organic Synthesis," p. 583, John Wiley & Sons, Inc., New York, 1967.
9. L. F. Fieser and M. Fieser, "Reagents for Organic Synthesis," p. 255, John Wiley & Sons, Inc., New York, 1967.
10. S. D. Gokhale and W. L. Jolly, *Inorganic Syntheses,* **9,** 56 (1957).
11. R. C. Marriott, J. D. Odom, and C. T. Sears, *ibid.,* **14,** 1 (1973).
12. A. D. Norman, *Inorg. Chem.,* **9,** 870 (1970).
13. K. M. Mackay, K. J. Sulton, S. R. Stobart, J. E. Drake, and C. Riddle, *Spectrochim. Acta,* **25A,** 925 (1969).

40. DIGERMYL SULFIDE

(Digermthiane)

$$2Li + S \longrightarrow Li_2S$$

$$Li_2S + 2GeH_3Br \longrightarrow (GeH_3)_2S + 2LiBr$$

Submitted by D. W. H. RANKIN*
Checked by J. M. BELLAMA† and L. A. HARMON†

Digermyl sulfide, prepared by the reaction of iodogermane with mercuric sulfide, was first reported in 1959.[1] More recently, it has been obtained from the reactions of germylphosphine or germylarsine with hydrogen sulfide[2] and by silent electric discharge on germane–hydrogen sulfide mixtures.[3] With the exception of the original method, the methods either give low yields or are indirect. The method described here involves the reaction of bromogermane and dilithium sulfide in dimethyl ether. High yields of digermyl sulfide (80%) can be obtained in about 10 hours; for most of this time the preparation does not require attention.

Digermyl sulfide has been used as an intermediate in the preparation of germanethiol (mercaptogermane),[2,4] and digermyl ether.[5] In view of the utility of disilyl sulfide as an intermediate,[6] it seems likely that the germyl analog will be exploited further.

Procedure

■ **Caution.** *The toxicity of digermyl sulfide is unknown. It should be handled in an efficient fume hood.*

The reactions are carried out by standard vacuum-line techniques[7] in an all-glass apparatus. Stopcocks may be greased, but the reaction itself is best carried out in a glass ampul with a greaseless stopcock.

Dilithium sulfide is first prepared. Ammonia (*ca.* 10 ml.) is condensed onto lithium (39 mg., 5.6 mmoles) and sulfur (93 mg., 2.9

*Department of Chemistry, University of Edinburgh, West Mains Road, Edinburgh EH9 3JJ, Scotland.
†Department of Chemistry, University of Maryland, College Park, Md. 20742.

mmoles) in a glass ampul (25–100 ml. volume) at $-196°$. The ampul is warmed to $-64°$ (chloroform slush) with shaking, until the blue color of dissolved lithium is completely discharged (*ca.* 1 hour). The dilithium sulfide produced is white. Excess lithium gives a deep-blue solution, whereas excess sulfur gives yellow or orange lithium polysulfides.

The ammonia is distilled out of the ampul, and the remaining solid is dried under vacuum for 2 hours to remove all traces of ammonia. Dimethyl ether (*ca.* 12 ml.) and bromogermane[8] (0.98 g., 6.3 mmoles, 10% excess) are then added, after which the ampul is maintained at $-96°$ (toluene slush) with shaking. Care should be taken to ensure that the bromogermane mixes with the solvent and does not remain as a solid on the walls of the reaction vessel.

After 6 hours the vessel is opened to the vacuum system and the contents are fractionated using traps maintained at $-23°$ (carbon tetrachloride slush), $-64°$, and $-96°$. The second trap retains digermyl sulfide (*ca.* 2.3 mmoles, 80% yield based on lithium). The separation of the product from bromogermane can be accomplished by distillation in a low-temperature column.[7]

The purity of the product is best checked by its infrared spectrum (2090, 2065, 931, 849, 819, 577, 556, 412 cm.$^{-1}$). Bromogermane has prominent bands at 2110, 870, 830, and 580 cm.$^{-1}$, and considerable rotation fine structure is visible on most of these. No such detail is visible on any of the bands in the spectrum of digermyl sulfide.

Properties

Digermyl sulfide is a colorless liquid with a vapor pressure of 1 mm. at $0°$, and an intense, nauseating odor. It is stable for short periods at room temperature in the absence of air and moisture but is best stored in sealed ampuls at liquid-nitrogen temperature.

The ^1H n.m.r. spectrum has a single line at $\tau 4.64$.

References

1. S. Sujishi, Abstracts, 17th International Congress of Pure and Applied Chemistry, 1959, p. 53.
2. J. E. Drake and C. Riddle, *J. Chem. Soc. (A),* **1968,** 2709.
3. J. E. Drake and C. Riddle, *ibid.,* **1970,** 3134.

4. C. Glidewell, D. W. H. Rankin, and G. M. Sheldrick, *Trans. Faraday Soc.,* **65,** 1409 (1969).
5. T. D. Goldfarb and S. Sujishi, *J. Am. Chem. Soc.,* **86,** 1679 (1964).
6. C. Glidewell, *J. Inorg. Nucl. Chem.,* **31,** 1303 (1969).
7. D. F. Shriver, "The Manipulation of Air-sensitive Compounds," McGraw-Hill Book Company, New York, 1969.
8. M. F. Swiniarski and M. Onyszchuk, *Inorganic Syntheses,* **15,** 157 (1974).

Chapter Seven

PHOSPHORUS COMPOUNDS

41. ETHYLENEBIS(DIMETHYLPHOSPHINE) [1,2-BIS(DIMETHYLPHOSPHINO)ETHANE], TETRAMETHYLDIPHOSPHANE DISULFIDE, AND TETRAMETHYLDIPHOSPHANE

Submitted by S. A. BUTTER* and J. CHATT†
Checked by E. R. WONCHOBA‡ and G. W. PARSHALL‡

Diphosphines are of considerable interest as ligands because of their excellent donor chelating properties with transition-metal ions, and are used to stabilize unusual oxidation states. Complexes formed with ethylenebis(dimethylphosphine) have increased electron density at the metal atom and are particularly reactive compared with the analogous ethylenebis(diphenylphosphine) complexes.[1]

The preparation of ethylenebis(dimethylphosphine) has been described by metallation of phosphine with sodium, followed by methylation of the resulting sodium phosphide with iodomethane. Two further metallation-methylation sequences using sodamide and iodomethane produce sodium dimethylphosphide, which is treated with 1,2-dichloroethane in liquid ammonia, to give the diphosphine.[2] Other procedures involve heating tetramethyldiphosphane with ethylene,[3] or tetramethyldiphosphane disulfide with ethylene, both

*Mobil Chemical Co., Research & Development Laboratories, Edison, N.J. 08817.
†University of Sussex, Brighton, England.
‡Central Research Department, E. I. du Pont de Nemours & Company, Wilmington, Del. 19898.

under pressure, the latter followed by desulfurization with tributyl-phosphine.[4]

The present procedure starts with tetramethyldiphosphane di-sulfide which is desulfurized with iron powder to tetramethyldiphos-phane. The diphosphane is cleaved with sodium in liquid ammonia, followed by addition of 1,2-dichloroethane, to give ethylenebis-(dimethylphosphine). This procedure avoids the use of phosphine (PH$_3$) and eliminates the need for high-pressure apparatus. Sections A and B for preparing tetramethyldiphosphane and the disulfide have been reported[5,6] and are reproduced with some modifications. The total working time for Secs. A, B, and C is 3 days.

A. TETRAMETHYLDIPHOSPHANE DISULFIDE*

$$Mg + CH_3I \longrightarrow CH_3MgI$$

$$6CH_3MgI + 2PSCl_3 \longrightarrow (CH_3)_2PS—PS(CH_3)_2 + 6MgClI + CH_3CH_3$$

Procedure

Methylmagnesium iodide[7] is prepared in a 3-l. flask containing 97 g. (4.0 g. atoms) of magnesium turnings in 750 ml. of sodium-dried ether. The flask is fitted with a nitrogen inlet, reflux condenser, thermometer, stirrer, and a 1-l. pressure-equalizing dropping funnel. A long-stem thermometer with the 0° region outside the flask is used, since deposits may obscure the markings on a short thermometer. A powerful motor must be used, since the mixture becomes viscous during the reaction with PSCl$_3$. Iodomethane (568 g., 4.0 moles) in 750 ml. of ether is added dropwise from the addition funnel at a rate sufficient to maintain reflux (*ca.* 5–6 hours addition period). Nearly all the magnesium should be consumed.

Under a stream of nitrogen, a solution of 217 g. (1.28 moles) of freshly distilled phosphorus trichloride sulfide† in 300 ml. of ether is added to the dropping funnel. The Grignard solution is cooled to

---

*The synthesis of tetramethyldiphosphane disulfide from commercially available methyl-magnesium bromide solution has been described in detail.[6] The disulfide is available from Strem Chemical Co., Danvers, Mass. 01923.

†Phosphorus trichloride sulfide (thiophosphoryl chloride) is available from Eastman Organic Chemicals or may be prepared from phosphorus trichloride.[8]

0° with a Dry Ice–acetone bath, and the phosphorus trichloride sulfide solution is added slowly (over 1.5 hours), maintaining the reaction temperature at 0° ± 5°. Some orange material may be noted during the addition, and the mixture becomes thick with white solids. When the addition is complete, the flask is allowed to warm slowly to room temperature, and is refluxed for 1.5 hours. The mixture is cooled to 0° with an ice-salt bath and poured into a 5-l. beaker one-half filled with cracked ice and 200 ml. of concentrated sulfuric acid. The residue in the flask is washed into the beaker with portions of ether and water. The product is allowed to stand overnight. The white solid is filtered, washed with ice water, then ether, and dried under vacuum. The yield of tetramethyldiphosphane disulfide is about 101 g. (85%), m.p. 224–228°. Recrystallization from 3:1 toluene-ethanol (27 ml./g. of crude) gives pure white needles, m.p. 235–237°. *Anal.* Calcd. for $C_4H_{12}P_2S_2$: C, 25.8; H, 6.5. Found: C, 25.8; H, 6.65.

Properties

Tetramethyldiphosphane disulfide is stable in air. The crystals are soluble in chloroform and acetone, moderately soluble in benzene and toluene, slightly soluble in alcohols, and insoluble in ether and pentane. In the solid state the preferred conformation is trans, as indicated by one strong P—S stretching absorption in the infrared spectrum (Nujol mull) at 567 cm.$^{-1}$, and the absence of a P—P band, which is found only in the Raman spectrum at 445 cm.$^{-1}$.[9] Other principal absorption bands are at 733, 745, 860, 882, 939, and 1281 cm.$^{-1}$.

B. TETRAMETHYLDIPHOSPHANE

$$(CH_3)_2PS—PS(CH_3)_2 + 2Fe \longrightarrow (CH_3)_2P—P(CH_3)_2 + 2FeS$$

Procedure

■ **Caution.** *In common with other alkyl diphosphanes, tetra-methyldiphosphane presents some hazard because of its toxicity and extreme susceptibility to air oxidation. Therefore, all operations must* *be performed in an efficient hood with a fire extinguisher at hand. All ground-glass joints should be thoroughly greased with a silicone lubricant.*

Thirty-nine grams (0.21 mole) of recrystallized tetramethyldiphosphane disulfide and 75 g. of iron powder (hydrogen-reduced or electrolytic) are mixed intimately in a 250-ml., single-neck, round-bottomed flask and degassed by evacuation and filling with nitrogen. The flask is fitted with a short Vigreux column (10–12 cm.), a cold-finger distillation head, and a previously tared 25- or 50-ml. pressure-equalizing dropping funnel as the receiver. (A standard distillation assembly with the dropping funnel under a fraction cutter may be used.) A Dry Ice trap is used after the receiver to condense any $(CH_3)_2PH$ which may form if the materials are not absolutely anhydrous. The complete apparatus is evacuated and filled with nitrogen two or three times, and the desulfurization is done under nitrogen. The flask is heated gently with a continuously moving Bunsen flame. The disulfide melts, and the mixture is refluxed gently for about 10 minutes. Then the flask is strongly heated to distill the colorless tetramethyldiphosphane, which is collected at 140–150°; yield is 22–23 g. (85–90%). This product is very flammable and should be kept in the dropping funnel under nitrogen until used in the following preparation.

C. ETHYLENEBIS(DIMETHYLPHOSPHINE)

[1,2-Bis(dimethylphosphino)ethane]

$$(CH_3)_2P\!-\!P(CH_3)_2 + 2Na \xrightarrow{NH_3(l)} 2NaP(CH_3)_2$$

$$2NaP(CH_3)_2 + ClCH_2CH_2Cl \longrightarrow (CH_3)_2PCH_2CH_2P(CH_3)_2 + 2NaCl$$

Procedure

■ **Caution.** *In common with other alkyl phosphines, ethylenebis-(dimethylphosphine) presents some hazard because of its toxicity and extreme susceptibility to air oxidation. Therefore, all operations must be performed in an efficient hood with a fire extinguisher at hand. Rubber gloves and face protection should be worn when handling liquid ammonia, sodium, and the phosphines. All ground-glass joints should be thoroughly greased with a silicone lubricant.*

A dry, nitrogen-flushed, two-necked, 500-ml. flask is fitted with a Dry Ice condenser connected to a nitrogen inlet. The second neck is used for the addition of sodium and liquid ammonia against a counterstream of nitrogen. A powerful magnetic stirring bar is placed in the flask for continuous stirring during the following operations. A mechanically sealed stirrer and gland may be used in a third neck, provided it does not leak during vacuum distillation. The flask is charged with 8.3 g. (0.36 g. atoms) of sodium chips,* and while cooling with a Dry Ice–acetone bath, 300–400 ml. of liquid ammonia are added. The dropping funnel containing 22 g. (0.18 mole) of tetra-methyldiphosphane is placed in the second neck. After about 0.5 hour of stirring, the diphosphane is added dropwise with cooling sufficient to prevent loss of ammonia from the reaction flask. (The temperature is satisfactory if the condenser returns all the ammonia to the flask.) The characteristic blue color of sodium in ammonia gradually changes to the red color of sodium dimethylphosphide during the 30 minutes addition period. The cooling bath is removed and 17.8 g. (0.18 mole)* of anhydrous 1,2-dichloroethane is cautiously added dropwise. The addition must be controlled so that the vigorous, exothermic reaction does not boil away the ammonia. The red color is gradually discharged, and sodium chloride precipitates. The mixture should be white at the end of the addition. Ammonia is allowed to evaporate under a slow stream of nitrogen until a white cake of salt moistened with the product remains.

(■ **Caution.** *Some toxic dimethylphosphine may be swept out with the ammonia.*) The nitrogen inlet is then placed on the flask in place of the condenser, and a short Vigreux column (10–12 cm.) replaces the dropping funnel. Also connected are an adapter with thermometer, a water condenser, and a receiver as shown in Fig. 6. The diphosphine is distilled directly from the salt cake (a nitrogen bleed is unnecessary) into the receiver under reduced pressure.

(■ **Caution.** *Dimethylphosphine, if formed by traces of moisture, is evolved and may collect in vacuum-line traps or may disperse through the sewer system if a water aspirator is used.*) The ethylenebis-(dimethylphosphine) distills at 65–66° at 10 torr; yield is 13.5 g. (50%).

---

*This amount should be modified relative to the yield of the diphosphane.

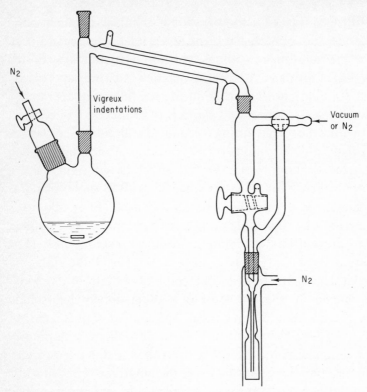

Fig. 6. Apparatus for filling ampuls with distillate.

The product is then sealed in ampuls under nitrogen using the apparatus shown in Fig. 6. This consists of a gradually constricted capillary tube (with a joint to fit the receiver) fitted into a test tube with the end cut off through which nitrogen is passed. A second stream of nitrogen is passed into the receiver and down through the ampul filler, thus excluding atmospheric oxygen. Ampuls, labeled and tared, dried, and nitrogen-flushed, are then filled and immediately sealed in a gas-air flame. Ampuls of 3 ml. capacity containing 0.5–2 ml. of diphosphine are most useful.

Properties

Ethylenebis(dimethylphosphine) is an air-sensitive, colorless liquid which boils at 180°/760 mm.; specific gravity, 0.90_4^{23}; n_D^{25}, 1.4887. It inflames on contact with air but may be kept indefinitely in sealed ampuls under nitrogen.

References

1. J. Chatt and S. A. Butter, *Chem. Commun.,* **1967**, 501; J. Chatt and J. M. Davidson, *J. Chem. Soc.,* **1965**, 843.
2. C. E. Wymore, Ph.D. thesis, University of Illinois, 1957: J. Chatt and R. G. Hayter, *J. Chem. Soc.,* **1961**, 896.
3. A. B. Burg, *J. Am. Chem. Soc.,* **83**, 2226 (1961).
4. G. W. Parshall, *J. Inorg. Nucl. Chem.,* **14**, 291 (1960).
5. M. I. Kabachnik and E. S. Shepeleva, *Izv. Akad. Nauk S.S.S.R.. Otd. Khim. Nauk,* **1949**, 56.
6. G. W. Parshall, *Org. Syn.,* **45**, 102 (1965).
7. G. O. Doak, G. G. Long, and M. E. Key, *Inorganic Syntheses,* **9**, 92 (1967).
8. T. Moeller, H. T. Birch, and N. C. Nielsen, *ibid.,* **4**, 71 (1953).
9. A. H. Cowley and W. D. White, *Spectrochim. Acta,* **22**, 1431 (1966).

42. DIMETHYLPHOSPHINOTHIOIC CHLORIDE AND DIMETHYLPHOSPHINOUS CHLORIDE

(*Chlorodimethylphosphine*)

Submitted by G. W. PARSHALL*
Checked by R. C. STOCKS† and L. D. QUIN†

Dialkylphosphinous chlorides (chlorodialkylphosphines) are valuable synthetic intermediates, but they have not generally been easily accessible.[1,2] Direct alkylation of PCl_3 with limited amounts of *n*-alkyl Grignard reagents usually gives mixtures of products, although the isopropyl[3] and *tert*-butyl[4] reagents cleanly give the corresponding dialkylphosphinous chlorides. Dimethylphosphinous chloride has been obtained in a single step from the conveniently available[5,‡] tetramethyldiphosphane disulfide by treatment with $C_6H_5PCl_2$[6] or $(C_6H_5)_2PCl$,[7] but these procedures give fairly low yields of pure $(CH_3)_2PCl$. A relatively clean synthesis of dimethylphosphinous chloride in two steps from tetramethyldiphosphane disulfide has been devised by Maier.[7] As described below, it involves cleavage of the P—P bond with SO_2Cl_2 (Cl_2 can also be used[8]) to

*Central Research Department, E. I. du Pont de Nemours & Company, Wilmington, Del. 19898.
†Department of Chemistry, Duke University, Durham, N.C. 27706.
‡Available from Strem Chemicals, Inc., 150 Andover Street, Danvers, Mass. 01923.

give $(CH_3)_2PSCl$, which is then desulfurized with tributylphosphine. As reported by Maier, the procedure is adaptable to the synthesis of both symmetric and asymmetric dialkylphosphinous chlorides.

A. DIMETHYLPHOSPHINOTHIOIC CHLORIDE

$$[(CH_3)_2PS]_2 + SO_2Cl_2 \longrightarrow 2(CH_3)_2PSCl + SO_2$$

Procedure

A suspension of 93 g. (0.50 mole) tetramethyldiphosphane disulfide[5] in 400 ml. of benzene in a 1000-ml. flask fitted with mechanical stirrer, thermometer, and addition funnel is stirred with cooling to keep the temperature at 20° while a solution of 41 ml. (0.51 mole) sulfuryl chloride in 100 ml. of benzene is added over a period of 40 minutes. The mixture, now containing little solid, is allowed to warm to room temperature, and the benzene is distilled through a Vigreaux column substituted for the addition funnel. After removal of benzene, the residue is distilled in a 40-cm. spinning-band column,* equipped with a calcium chloride tube to exclude moisture, to give dimethyl-phosphinothioic chloride, b.p. 70–72° at 12 torr, yield 111 g. (86%). The product is a colorless, highly refractive liquid (n_D^{25}, 1.5392) with a freezing point above 0°. It is slightly moisture-sensitive and should be stored in sealed containers.

B. DIMETHYLPHOSPHINOUS CHLORIDE
(Chlorodimethylphosphine)

$$(CH_3)_2PSCl + (n\text{-}C_4H_9)_3P \longrightarrow (CH_3)_2PCl + (n\text{-}C_4H_9)_3PS$$

Procedure

■ **Caution.** *Dimethylphosphinous chloride is toxic and spontaneously flammable. All operations should be carried out in an efficient fume hood with careful exclusion of air from the product.*

The flask of a 40-cm. spinning-band column† fitted for collection

---

*The checkers found that the distillation could be done with a 30-cm. Vigreux column with no loss in quality. Their yield was 88%.

†The checkers used a 30-cm. Vigreux column for the reaction with good success. A little $(CH_3)_2PSCl$ co-distilled and appeared as a heavy oily layer in the crude distillate. It was easily separated by decantation.

of distillate under nitrogen* is charged with 116 g. of dimethyl-phosphinothioic chloride and 233 ml. of tri-n-butylphosphine. The two liquids become completely miscible at 110°, and reaction with evolution of dimethylphosphinous chloride begins at a pot temperature of 125°. The crude product distills at *ca.* 72°. Redistillation from 5 ml. of fresh tri-n-butylphosphine gives 63.5 g. (73% yield) of clear, colorless dimethylphosphinous chloride, b.p. 73° at 750 torr.

Properties

Dimethylphosphinous chloride is exceedingly air-sensitive and is best stored under vacuum or N_2. It melts at -1.4 to $-1.0°$, and its vapor pressure is given by the equations:[9] $\log_{10}p = 12.1408 - 2887/T$ (solid) and $\log_{10}p = 7.844 - 1722/T$. The ^{31}P n.m.r. signal has a chemical shift of -93.0 p.p.m. relative to external 85% H_3PO_4.

References

1. G. M. Kosolapoff, "Organophosphorus Compounds," John Wiley & Sons. Inc., New York, 1950.
2. K. Sasse, in E. Müller, "Methoden der Organischen Chemie" (Houben-Weyl), 4th ed., Vol. 12, No. 1, p. 203, Georg Thieme Verlag, Stuttgart, 1963.
3. W. Voskuil and J. F. Arens, *Org. Syn.*, **48**, 47 (1968).
4. M. Fild, O. Stelzer, and R. Schmutzler, *Inorganic Syntheses*, **14**, 4 (1973).
5. G. W. Parshall, *Org. Syn.*, **45**, 102 (1965); S. A. Butter and J. Chatt, *Inorganic Syntheses*, **15**, 185 (1974).
6. G. W. Parshall, *J. Inorg. Nucl. Chem.*, **12**, 372 (1960).
7. L. Maier, *Chem. Ber.*, **94**, 3051 (1961).
8. W. Kuchen and H. Buchwald, *Angew. Chem.*, **71**, 162 (1959).
9. A. B. Burg and P. J. Slota, *J. Am. Chem. Soc.*, **80**, 1107 (1958).

---

*A suitable apparatus is described in *Inorganic Syntheses*. **11**, 158 (1968) and on p. 190 of this volume.

43. PHOSPHORIC TRIHALIDES

(*Phosphoryl Halides*)

Submitted by HERBERT W. ROESKY*
Checked by W. STADELMANN,† O. STELZER,† and R. SCHMUTZLER †

The syntheses of phosphoric chloride difluoride and dichloride fluoride have been previously carried out by fluorination of phosphoryl trichloride with NH_4F, SbF_3, CaF_2, AgF, HF, NaF, potassium fluorosulfinate, or benzoyl fluoride under carefully controlled conditions.[1-8] The phosphoryl fluorohalides were obtained as by-products with POF_3 predominating. The reaction of calcium fluoride, sodium chloride, and phosphorus(V) oxide produces both $POFCl_2$ and POF_2Cl.[9] The cleavage of the P—N bond in $[(CH_3)_2N]_2POF$ with hydrogen chloride yields $POFCl_2$.[10] In the present method phosphorodifluoridic acid(difluorophosphoric acid) is allowed to react with phosphorus pentachloride to produce POF_2Cl in nearly quantitative yield. The only volatile products are POF_2Cl and HCl. Phosphoric chloride difluoride free of HCl may be obtained by first forming the anhydride of phosphorodifluoridic acid (diphosphoric tetrafluoride), which is then allowed to react with phosphorus pentachloride.[11]

Phosphoric dichloride fluoride (fluorophosphoryl dichloride) can be prepared more conveniently by treating phosphorus pentachloride with phosphorofluoridic acid (monofluorophosphoric acid). Under these conditions $POFCl_2$ is the only fluorinated product.[12] Similarly POF_2Br and $POFBr_2$ can be synthesized by partial fluorination of $POBr_3$.[13,14] Bromination and chlorination of $C_4H_9OPF_2$ in toluene at low temperatures have been reported to give POF_2Br and POF_2Cl.[15] The methods described below involving the reactions of POF_2OH or of $P_2O_3F_4$ with phosphoric pentabromide are convenient for the preparation of POF_2Br on a large scale.

*Anorganisch-Chemisches Institut der Universität Frankfurt, Robert-Mayer-Strasse 7-9, 6 Frankfurt/Main, West Germany.

†Lehrstuhl B für Anorganische Chemie der Technischen Universität, Pockelsstrasse 4, 33 Braunschweig, West Germany.

The phosphoryl fluorides are a source of many inorganic and organic derivatives.[16,17] The easy cleavage of a phosphorus-chlorine or phosphorus-bromine bond is responsible for the synthetic utility of the compounds.

■ **Caution.** *Although the toxicity of the phosphoryl fluorohalides has not been investigated in detail, precautions similar to those suggested for the handling of volatile phosphorus-fluorine compounds should be taken. These preparations should be performed only in a well-ventilated hood, and contact with the skin should be avoided. Because the compounds hydrolyze easily, they are vesicants like hydrogen fluoride.*

A. PHOSPHORIC CHLORIDE DIFLUORIDE
(Difluorophosphoryl Chloride)

Procedure

1. From Phosphorodifluoridic Acid (Difluorophosphoric Acid)

$$HPO_2F_2 + PCl_5 \longrightarrow POF_2Cl + POCl_3$$

A three-necked 250-ml. flask fitted with a reflux condenser, dropping funnel, and mechanical stirrer is charged with 25 g. (0.12 mole) of phosphorus pentachloride. Two cold traps are attached to the outlet of the condenser. The first trap is cooled with Dry Ice–acetone and the second trap with liquid nitrogen.* The dropping funnel is charged with 10 g. (0.10 mole) of phosphorodifluoridic acid† (difluorophosphoric acid), which is added slowly to the phosphorus pentachloride. During the course of the addition POF_2Cl and some hydrogen chloride are collected in the first trap while most of the hydrogen chloride passes the first trap and is collected in the second trap. When the addition of the phosphorodifluoridic acid is complete, the reaction mixture is stirred for 2 hours and warmed slowly to 60° to remove dissolved POF_2Cl. In order to remove the hydrogen chloride from the first trap, the crude product is condensed a second time with two new cold traps, keeping the first at about -25 to $-30°$ and cooling the second with liquid nitrogen.* The trap with the crude product is warmed slowly to room temperature. The

---

* ■ **Caution.** A nitrogen atmosphere must be maintained throughout the system to avoid condensation of atmospheric oxygen in these traps.

†Available from Ozark-Mahoning Co., 1870 South Boulder, Tulsa, Okla. 74119.

yield is 10–11 g. (79%). The reaction may be conveniently scaled up for the preparation of larger quantities.

2. From Diphosphoric Tetrafluoride (Phosphorodifluoridic Anhydride, Difluoro-phosphoric Acid Anhydride)

$$P_2O_3F_4 + PCl_5 \longrightarrow 2\ POF_2Cl + POCl_3$$

The diphosphoric tetrafluoride for this synthesis is best prepared[18] by reaction of phosphorodifluoridic acid with an excess of phosphorus(V) oxide. Extremely pure $P_2O_3F_4$ is not required. It contains about 5% of the starting material. The same procedure, but using 18.6 g. (0.1 mole) of diphosphoric tetrafluoride instead of the acid in the method given above, may be utilized to prepare POF_2Cl. The purification step described above is not necessary in this synthesis.[12] The yield is 23 g. (95%).

Properties

Phosphoric chloride difluoride (difluorophosphoryl chloride) is a colorless gas which is readily hydrolyzed by atmospheric moisture, melts at $-96.4°$, and boils at $3.1°$. The vapor pressure is expressed by the following equation: $\log p = -1328.3/T + 7.6904$. The liquid density is 1.6555 at $0°$. The ^{31}P n.m.r. spectrum shows a triplet (15 p.p.m. vs. 85% H_3PO_4). The $^{19}F—^{31}P$ coupling constant varies from 1120 to 1145 Hz., according to different authors.[19-23]

B. PHOSPHORIC DICHLORIDE FLUORIDE

(Fluorophosphoryl Dichloride)

$$POF(OH)_2 + 2\ PCl_5 \longrightarrow POFCl_2 + 2\ POCl_3 + 2\ HCl$$

Procedure

In a three-necked 1-l. flask equipped with a dropping funnel, high-capacity reflux condenser and mechanical stirrer are placed 208 g. (1.0 mole) of phosphorus pentachloride. A T-shaped adapter is attached to the outlet of the condenser, which is flushed with dry nitrogen to prevent moisture from entering the reaction vessel. Fifty grams (0.5 mole) of phosphorofluoridic acid* (monofluorophosphoric

---

*Available from Ozark-Mahoning Co., Tulsa, Okla.

acid) are added dropwise to the phosphorus pentachloride. The hydrogen chloride which is formed during the reaction is allowed to escape in the hood. After the addition is complete, the reflux condenser is replaced by a distillation head, and the reaction flask is heated gently till the temperature at the distillation head reaches 100°. The crude product is redistilled using a 60-cm. vacuum-jacketed column filled with glass helices. Phosphoric dichloride fluoride distills at 53–54° at 760 torr; yield is 27–41 g. (40–60%).*

Properties

Phosphoric dichloride fluoride (fluorophosphoryl dichloride) is a colorless liquid which boils between 52.2 and 54° and melts at $- 80.1$ or $- 84.5°$. It has a liquid density of 1.5931 at 0°. The vapor pressure is expressed by the following equation: $\log p = - 1618.2/T + 7.8440$. The P—F distance was found to be 1.50 ± 0.03 A. and the P—O 1.54 ± 0.03 A., compared with 1.51 ± 0.03 A. and 1.55 ± 0.03 A. in POF_2Cl. The phosphorus-chlorine distance[24] is slightly different, 1.99 ± 0.04 A. in $POFCl_2$ and 2.01 ± 0.04 A. in POF_2Cl. The FPF angle in POF_2Cl and the ClPCl angle in $POFCl_2$ are equal to $106 \pm 0.3°$. The ^{31}P n.m.r. spectrum shows a doublet centered at 0.0 or $- 0.8$ p.p.m. (vs. 85% H_3PO_4). The ^{31}P—^{19}F coupling constant was found to be 1175 or 1190 Hz.[25]

C. PHOSPHORIC BROMIDE DIFLUORIDE

(Difluorophosphoryl Bromide)

$$POF_2OH + PBr_5 \longrightarrow POF_2Br + POBr_3 + HBr$$

Procedure

Phosphorodifluoridic acid (difluorophosphoric acid) (51 g., 0.5 mole) and 258 g. (0.6 mole) of phosphorus pentabromide are placed in a two-necked 500-ml. flask equipped with a distillation head and a mechanical stirrer. The reaction flask is heated in an oil bath within 2 hours to 90–95°, and heating is continued for another 2 hours at 95–105°. The resulting volatile liquid is distilled during

---

* This depends on the quality of the phosphorofluoridic acid.

this procedure, and the collecting flask is cooled with wet ice. The checkers found it necessary to heat to 160°. The crude product is redistilled using a 25-cm. vacuum-jacketed column filled with glass helices. Phosphoric bromide difluoride distills at 28–30° at 760 torr; yield is 20 g. (60 %). The product is contaminated by small amounts of elemental bromine.

Properties

Phosphoric bromide difluoride (difluorophosphoryl bromide) boils at 30.5° and melts at $-84°$. The liquid density is 2.099 at 0°. Its vapor pressure follows the equation: $\log p = -1550.0/T + 7.9662$. The ^{31}P n.m.r. spectrum shows a triplet with a $^{31}P-^{19}F$ coupling constant of 1203 Hz.

References

1. C. J. Wilkins, *J. Chem. Soc.*, **1951**, 2726.
2. H. S. Booth and F. B. Dutton, *J. Am. Chem. Soc.*, **61**, 2937 (1939).
3. H. Moissan, *Bull. Soc. Chim. Paris*, [3] **5**, 456 (1891).
4. K. Wiechert, *Z. Anorg. Chem.*, **261**, 310 (1950).
5. F. Seel and L. Riehl, *ibid.*, **282**, 293 (1955).
6. F. Seel, K. Ballreich, and W. Peters, *Chem. Ber.*, **92**, 2117 (1959).
7. C. W. Tullock and D. D. Coffmann, *J. Org. Chem.*, **25**, 2016 (1960).
8. N. B. Chapman and B. C. Saunders, *J. Chem. Soc.*, **1948**, 1010.
9. G. Tarbutton, E. P. Egan, Jr., and S. G. Frary, *J. Am. Chem. Soc.*, **63**, 1782 (1941).
10. Z. Skrowaczewska and P. Mastalerz, *Roczniki Chem.*, **29**, 415 (1955).
11. H. W. Roesky, *Chem. Ber.*, **101**, 636 (1968).
12. H. W. Roesky, *Z. Naturforsch.*, **24b**, 818 (1969); E. Fluck and W. Steck, *Syn. Inorg. Metal-org. Chem.*, **1**, 29 (1971).
13. H. S. Booth and C. G. Seegmiller, *J. Am. Chem. Soc.*, **61**, 3120 (1939).
14. G. A. Olah and A. A. Oswald, *J. Org. Chem.*, **24**, 1568 (1959).
15. G. I. Drozd and S. Z. Ivin, *Zh. Obshch. Khim.*, **38**, 1907 (1968).
16. R. Schmutzler, Fluorides of Phosphorus, in "Advances in Fluorine Chemistry," Butterworth & Co. (Publishers), Ltd., London, 1965.
17. H. W. Roesky and E. Niecke, *Z. Naturforsch.*, **24b**, 1101 (1969).
18. E. A. Robinson, *Can. J. Chem.*, **40**, 1725 (1962).
19. H. S. Gutowsky and D. W. McCall, *Phys. Rev.*, **82**, 748 (1951).
20. H. S. Gutowsky, D. W. McCall, and C. P. Slichter, *J. Chem. Phys.*, **21**, 279 (1953).
21. H. S. Gutowsky and D. W. McCall, *ibid.*, **22**, 162 (1954).
22. R. A. Y. Jones and A. R. Katritzky, *Angew. Chem.*, **74**, 60 (1962).
23. R. R. Holmes and W. P. Gallagher, *Inorg. Chem.*, **2**, 433 (1963).
24. L. O. Brockway and J. Y. Beach, *J. Am. Chem. Soc.*, **60**, 1836 (1938).
25. M. M. Crutchfield, C. F. Callis, and J. R. Van Wazer, *Inorg. Chem.*, **3**, 280 (1964).

44. μ-NITRIDO-BIS[AMIDODIPHENYLPHOSPHORUS](1+) CHLORIDE

$$(C_6H_5)_2PCl + Cl_2 \longrightarrow (C_6H_5)_2PCl_3$$

$$2(C_6H_5)_2PCl_3 + 8NH_3 \longrightarrow [(C_6H_5)_2P\text{===}N\text{===}P(C_6H_5)_2]^+Cl^- + 5NH_4Cl$$
$$\underset{NH_2}{|} \qquad \underset{NH_2}{|}$$

Submitted by MANFRED BERMANN* and JOHN R. VAN WAZER*
Checked by J. K. RUFF† and J. ZINICH†

μ-Nitrido-bis[amidodiphenylphosphorus](1 +) chloride may be obtained (1) by ammonolysis of trichlorodiphenylphosphorane,[1] (2) by chloramination of chlorodiphenylphosphine,[2] tetraphenyl-diphosphine,[3] or iminobis (diphenylphosphine),[4] (3) by ammonolysis of μ-nitrido-bis[chlorodiphenylphosphorus] (1 +) chloride,[5] or (4) as a by-product of the ammonolysis of $(PNCl_2)_3$ followed by reaction with $(C_6H_5)_2PCl_3$.[6]

The most convenient way combined with highest yields is the direct ammonolysis of $(C_6H_5)_2PCl_3$ with gaseous ammonia, as described in the following procedure. Ammonolysis with *in situ* produced NH_4Cl[1] is complicated; chlorination and ammonolysis in one step[7] [i.e., without the isolation of $(C_6H_5)_2PCl_3$] is awkward and touchy. The chloramination methods[2-4] necessitate an efficient chloramine generator and furthermore do not permit the synthesis of large quantities.

Scaling up of the procedure given herein is not recommended, because handling becomes rather difficult. Other diaryltrichloro-phosphoranes may be similarly subjected to ammonolysis, but the method fails with dialkyltrichlorophosphoranes.[5] A total elapsed time of 2 days should be allowed for carrying out this over-all preparation at the scale given. μ-Nitrido-bis[amidodiphenylphosphorus](1 +) chloride is a useful chemical intermediate, particularly for specific ring-closure reactions as described in the Properties section.

* Department of Chemistry, Vanderbilt University, Nashville, Tenn. 37235.
† Department of Chemistry, University of Georgia, Athens, Ga. 30601.

Procedure

Preparation of Trichlorodiphenylphosphorane. In a 2-l., three-necked, round-bottomed flask equipped with a paddle stirrer, a condenser with a $CaCl_2$ tube, and a gas inlet tube, and previously flushed with nitrogen, is placed 220.5 g. (1 mole) of $(C_6H_5)_2PCl^*$ in 1400 ml. of carbon tetrachloride (ACS Reagent or Spectro grade). The system is cooled with an NaCl–ice bath to *ca.* 0°, and dry chlorine (Matheson Co.) is passed *over* the solution. A white precipitate of $(C_6H_5)_2PCl_3$ forms immediately. The chlorine is passed over the solution until the color of the precipitate turns slightly yellow (2 hours) because of excess chlorine which is absorbed; this excess of chlorine cannot be removed later. A few milliliters of $(C_6H_5)_2PCl$ are added until decoloration is achieved.† The extremely hygroscopic $(C_6H_5)_2PCl_3$ is filtered in a glove box and dried *in vacuo*, conveniently, in the three-necked flask which was used in its preparation. This flask may then be used in the second step. The yield of $(C_6H_5)_2PCl_3$ is practically quantitative. The white crystalline product is pure; m.p. (capillary) = 194–197°C. (decomposes).

Ammonolysis of Trichlorodiphenylphosphorane. The previously employed apparatus is used again, the drying tube being replaced by a NaOH-filled tube. Trichlorodiphenylphosphorane (281.9 g.) is dissolved in 1.5 l. of chloroform. Anhydrous ammonia‡ is introduced via a large inlet tube *into* the solution with vigorous stirring while maintaining the temperature between 0 and 5°. Temperature control is essential as higher temperatures give noticeable amounts of $[(C_6H_5)_2PN]_4$ and thus lower the yield of $\{[(C_6H_5)_2(H_2N)P]_2N\}Cl$. This operation requires approximately 2–3 hours. After the addition is complete, the heterogeneous mixture is filtered through a sintered-glass funnel and air-dried,§ leaving 249.5 g. of a mixture of NH_4Cl and $\{[(C_6H_5)_2(H_2N)P]_2N\}$ Cl. The solid material is washed once

---

*$(C_6H_5)_2PCl$ may be obtained from Stauffer Chemical Corp., Orgmet, Inc., Alfa Inorganics, and other sources and does not need to be purified by distillation.

†If the excess chlorine is not removed, chloramine will be formed during the ammonolysis, and yields are drastically reduced.

‡Passage of ammonia through a tube 30 cm. long and 3 cm. in diameter containing NaOH pellets gives appropriately anhydrous ammonia.

§A sintered-glass funnel is best, as the precipitate is very fine. Alternatively, if a sufficiently large funnel of this type is not available, a Büchner funnel may be used; but then the heterogeneous precipitate must be dried *in vacuo*.

with 400 ml. and three times with 200 ml. of cold water to remove the ammonium chloride. It is washed with ether and finally air-dried to give 120.5 g. of crude $\{[(C_6H_5)_2(H_2N)P]_2N\}Cl$.

The chloroform filtrate is concentrated to about 300 ml., and excess ether (600 ml.) is added and filtered to give 51.3 g. of crude $\{[(C_6H_5)_2(H_2N)P]_2N\}Cl$. The filtrate is evaporated to dryness, giving an oil which in turn is treated with 400 ml. of ether and allowed to stand overnight in the refrigerator. This procedure yields another 13.9 g. of $\{[(C_6H_5)_2(H_2N)P]_2N\}Cl$.

The combined fractions are recrystallized from 300 ml. of methanol to give the pure product as white, shiny needles, m.p. 245–246° (decomposes), at a total yield of $\{[(C_6H_5)_2(H_2N)P]_2N\}Cl$ of 178.7 g. (82 % of theory) for thorough recrystallization. The checkers, operating on one-tenth this scale, obtained a 49 % yield of recrystallized product, m.p. 244–246°. They found it necessary to add a little ether to the methanol to induce crystallization. *Anal.* Calcd.: C, 69.1; H, 5.76; N, 10.9. Found: C, 68.9; H, 5.87; N, 10.7.

Properties

μ-Nitrido-bis[amidodiphenylphosphorus](1 +) chloride is a nonhygroscopic, white solid (colorless in larger crystals). Crystallization from chloroform or acetonitrile gives 1:1 adducts, both of which are destroyed either by recrystallization from methanol or less completely by drying *in vacuo* at *ca.* 80° for several hours. The compound is very soluble in methanol, ethanol, dimethyl sulfoxide, dimethylformamide, and dimethylacetamide, exhibits moderately good solubility in chloroform, dichloromethane, and acetonitrile, and is practically insoluble in all other common organic solvents.

Electric-conductivity measurements in acetonitrile show it to be a 1:1 (associated) electrolyte.[5,8] The crystal structure has been determined.[9]

N.m.r.: Proton spectrum (DMSO-d_6, TMS): $\tau C_6H_5 = 1.98$, 2.20, 2.38 (complex multiplet); $\tau NH_2 = 3.87$.

Phosphorus spectrum (in CH_3OH; vs. 85 % H_3PO_4 ext.): $\delta_P = -18.0$ p.p.m.

Infrared (KBr, 0.5%): 3160–3140 (vs), 3060–3040 (vs), 1590 (w), 1553 (m), 1482 (m), 1438 (vs), 1260–1230 (vs), 1180 (s), 1160 (w), 1120–

1110 (vs), 1070 (w), 995 (s), 963 (vs), 933 (vs), 840 (w), 795 (vs), 715 (vs), 685 (vs), 610 (m, sh), 530 (vs), 510 (vs), 448 (m), 422 (m) cm.$^{-1}$.

References

1. I. I. Bezman and J. H. Smalley, *Chem. Ind.,* **1960**, 839; U.S. Patent 3,080,422 (1963) (Armstrong Cork Co.); *Chem. Abstr.,* **59,** 8790 (1963); I. I. Bezman, U.S. Patent 3,098,871 (1963) (Armstrong Cork Co.); *Chem. Abstr.,* **59,** 14024 (1963); R. G. Rice and B. Grushkin, U.S. Patent 3,329,716 (1967) (W. R. Grace and Co.); German Patent 1,222,500 (1966); British Patent 1,016,467 (1966); *Chem. Abstr.,* **64,** 17639 (1966).
2. H. H. Sisler, H. S. Ahuja, and N. L. Smith, *Inorg. Chem.,* **1,** 84 (1962); H. H. Sisler, German Patent 1,189,077 (1965) (W. R. Grace and Co.); British Patent 974,603 (1964); French Patent 1,339,384 (1963); *Chem. Abstr.,* **60,** 3012 (1964); U.S. Patent 3,418,366 (1968); I. I. Gilson and H. H. Sisler, *Inorg. Chem.,* **4,** 273 (1965).
3. S. E. Frazier and H. H. Sisler, *Inorg. Chem.,* **5,** 925 (1966).
4. D. F. Clemens and H. H. Sisler, *ibid.,* **4,** 1222 (1965); D. F. Clemens, M. L. Caspar. D. Rosenthal, and R. Peluso, *ibid.,* **9,** 960 (1970).
5. M. Bermann, unpublished results.
6. R. Keat, M. C. Miller, and R. A. Shaw, *J. Chem. Soc.* (*A*), **1967,** 1404.
7. C. Derderian, thesis, Pennsylvania State University, University Park, Pa., 1966.
8. I. Y. Ahmed and C. D. Schmulbach, *J. Phys. Chem.,* **71,** 2358 (1967).
9. J. W. Cox and E. R. Corey, *Chem. Commun.,* **1969,** 205.

Chapter Eight

MAIN GROUP AND ACTINIDE COMPOUNDS

45. TRIMETHYLGALLIUM

$$GaCl_3 + 3(CH_3)_3Al \longrightarrow Ga(CH_3)_3 + 3(CH_3)_2AlCl$$

Submitted by D. F. GAINES,* JORJAN BORLIN,* and E. P. FODY*
Checked by J. P. OLIVER† and A. SADURSKI†

Previously published syntheses of trimethylgallium, a catalyst in the polymerization of ethylene, include the reaction of gallium metal with dimethylmercury over a period of 3 days[1] and the sealed-tube reaction between gallium trichloride and dimethylzinc.[2,3]

A superior method involves the reaction of gallium trichloride with excess neat trimethylaluminum.[4,5] The trimethylgallium is separated from the reaction mixture by distillation and purified further by distillation from sodium fluoride in order to remove any chlorodimethylaluminum. This synthesis produces a high yield of trimethylgallium (*ca.* 63%) in about 6 hours and may be run on any convenient scale.

Procedure

■ **Caution.** *Trimethylaluminum, trimethylgallium, and chloro-dimethylaluminum burn spontaneously in air and react violently with*

*Department of Chemistry, University of Wisconsin, Madison, Wis. 53706.
†Department of Chemistry, Wayne State University, Detroit, Mich. 48202.

water. This reaction should be carried out in a hood, and personnel should wear fireproof face shields, gloves, and aprons. A sodium hydrogen carbonate or similar dry-powder fire extinguisher should be immediately available.

Sodium fluoride is dried *in vacuo* overnight. The system (Fig. 7), containing 10 g. of sodium fluoride in flask *D*, is flushed with dry nitrogen for at least one hour.* A glass ampul containing 25 g. of gallium trichloride† is placed inside a small plastic bag which has a wall thickness of at least 0.002 in. (Alternatively two bags, one inside the other, afford added protection against hydrolysis.) The bag is purged with dry nitrogen, and the ampul is subsequently crushed

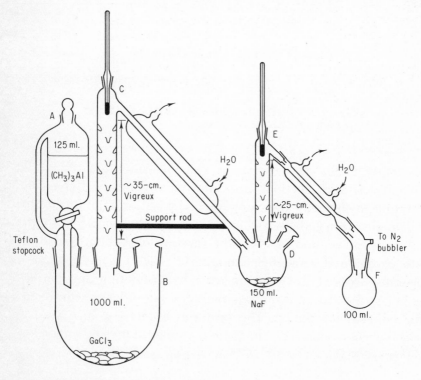

Fig. 7. *Apparatus for trimethylgallium synthesis.*

---

*AP 101 grease, obtained from James G. Biddle Co., Plymouth Meeting, Pa., was used on all joints; Teflon joint liners were also employed successfully. Other greases are not recommended. The addition funnel should be equipped with a Teflon stopcock plug to minimize the possibility of seizing while in contact with trimethylaluminum.

†The gallium trichloride was obtained from Eagle-Picher Industries, Quapaw, Okla. 74363.

with a hammer. Care should be taken that the broken glass does not puncture the bag (in practice this gave little trouble). The plastic bag is placed in a large nitrogen-filled glove bag and its contents, including broken glass, are emptied into the reaction flask *B*. The flask is attached quickly to the rest of the reaction apparatus, and about 80 ml. of trimethylaluminum* (a twofold excess) is transferred from a lecture cylinder to the addition funnel. This transfer is accomplished most readily by inverting the lecture bottle, attaching a hose connector (to which a 4-in. length of *ca.* 1/8-in.-o.d. stainless-steel tubing is soldered), and cautiously opening the main lecture-bottle valve *after* the stainless-steel tube has been inserted into the addition funnel.

The trimethylaluminum is added slowly to the gallium trichloride over a period of about one hour. The reaction is extremely exothermic and is controlled by varying the addition rate of trimethylaluminum. External cooling should be avoided, as this may adversely affect the product yield. Initially, a few drops at a time are added; later, the rate may be increased. The flask contents are stirred magnetically during the reaction and distillation.

Immediately after the addition of trimethylaluminum is completed, external heating is begun and the crude product is distilled at 55–60° onto the sodium fluoride in flask *D*. The pure product is redistilled at about 56° from sodium fluoride into the receiving flask *F*. This flask is taken from the system, quickly fitted with an adapter equipped with a stopcock, and attached to a vacuum line for transfer into a storage vessel. (■ **Caution.** *Some flaming of the product may occur when the receiving flask is removed from the apparatus*; however, rapid and careful work should prevent any serious problems.) Extreme caution must be exercised in the disposal of the pyrophoric residues in flask *B*.

The pyrophoric residues in flask *B* may be safely disposed of using the following procedure. After flask *B* has cooled to room temperature, about 900 ml. of heptane is run into it through the addition funnel *A*. The mixture is then stirred magnetically to ensure that all soluble residues are dissolved. The resulting 10%

---

*Trimethylaluminum may be purchased from the Ethyl Corporation, Baton Rouge, La., or from Texas Alkyls, P.O. Box 600, Deer Park, Tex. 77536.

solution in heptane does not appear to be pyrophoric (it does smoke, however, when exposed to air). Flask *B* is then removed from the rest of the apparatus and its contents slowly poured onto 2–3 lb. of crushed Dry Ice contained in a 5-gal. metal bucket. After the Dry Ice–heptane solution slurry has stood for about 0.5 hour, 1 l. of 95 % ethyl alcohol is added slowly to ensure complete solvolysis. The bucket is then left undisturbed until it has warmed to room temperature.

An alternate method suggested by the checkers is as follows. After flask *B* has cooled to room temperature about 600 ml. of a 10 % ethyl acetate solution in heptane is run slowly into it through the addition funnel *A*. This addition is followed by slow addition of 300 ml. of 30 % ethyl acetate in heptane. The mixture is stirred magnetically until all visible reaction has ceased. The resulting solution is cautiously poured onto 5 lb. of crushed ice in a 5-gal. metal bucket to complete the hydrolysis.

Care should be taken to dilute any remaining residues with heptane as the apparatus is being disassembled for cleaning. Any solid residues remaining in the apparatus should be covered with heptane and deactivated by slow addition of pentyl alcohol until there is no further evidence of reaction.

Properties[6]

Trimethylgallium is a clear, colorless liquid that melts at -15.7 to $-15.9°$ and boils at $55.7°$. It is very reactive toward oxygen and water but may be stored indefinitely under dry nitrogen or its own vapor pressure. It is a strong Lewis acid, forming many stable complexes; its chemical and physical properties have been well studied.[7,8]

References

1. G. E. Coates, *J. Chem. Soc.*, **1951**, 2003.
2. Wiberg, Johannsen, and Stecher, *Z. Anorg. Chem.*, **251**, 114 (1943).
3. L. I. Zakharkin and O. Yu. Okhlobystin, *Zh. Obshch. Khim.*, **31**, 3662 (1961); *Chem. Abstr.*, **57**, 8593c (1962).

4. Siemens-Schukerwerke, A. -G., German Patent 1,158,977 (1963); *Chem. Abstr.,* **60**, 6867h (1964).
5. G. E. Coates and K. Wade, "Organometallic Compounds, The Main Group Elements," Methuen & Co., Ltd., London, 1967.
6. C. A. Kraus and F. E. Toonder, *Proc. Natl. Acad. Sci.,* **19**, 292 (1933).
7. J. R. Hall, L. A. Woodward, and E. A. V. Ebsworth, *Spectrochim. Acta,* **20**, 1249 (1964).
8. K. Yasuda and R. Okaware, *Organometallic Chem. Rev.,* **2**, 255–277 (1967).

46. HEXAMETHYLDISILTHIANE AND HEXAMETHYLCYCLOTRISILTHIANE

$$[(CH_3)_3Si]_2S \text{ and } [(CH_3)_2SiS]_3$$

Submitted by D. A. ARMITAGE,* M. J. CLARK,† A. W. SINDEN,* J. N. WINGFIELD,‡
E. W. ABEL§ (Secs. B and C), and E. J. LOUIS¶ (Sec. A)
Checked by E. J. LOUIS¶ (Secs. B and C) and D. A. ARMITAGE* (Sec. A)

The methods for the preparation of hexamethyldisilthiane that appear in the literature suffer from inconvenience, high cost of materials, and low yields, or they require equipment not ordinarily available in the laboratory.[1-6] Two methods of considerably greater convenience have been developed and are described in detail here.

The reaction between hydrogen sulfide and 1-(trimethylsilyl)-imidazole[7] in the absence of solvent cleanly gives hexamethyl-disilthiane as described in Sec. A. An alternate method employs the reaction of chlorotrimethylsilane with hydrogen sulfide in the presence of a tertiary amine (Sec. B). This latter procedure was developed from a previously reported[8] synthesis of hexamethylcyclo-trisilthiane, described in more detail in Sec. C. Both silthianes are useful synthetic intermediates because they avoid the inconvenience of handling anhydrous sulfides or hydrogen sulfide.[9]

*Department of Chemistry, Queen Elizabeth College, London, W8 7AH, England.
†School of Chemical Sciences, University of East Anglia, Norwich NOR88C, England.
‡Department of Chemistry, Kings College, The Strand, London WC2, England.
§Department of Chemistry, The University, Exeter EX44QD, England.
¶Department of Chemistry, Lehman College of CUNY, New York, N.Y. 10021.

A. HEXAMETHYLDISILTHIANE VIA 1-(TRIMETHYLSILYL)-IMIDAZOLE

1. Trimethylsilylimidazole[10]

$$[(CH_3)_3Si]_2NH + 2HN \underset{\diagdown}{\overset{=N}{\diagup}} \longrightarrow 2(CH_3)_3Si-N \underset{\diagdown}{\overset{=N}{\diagup}} + NH_3$$

Procedure

■ **Caution.** *The Si—N bonded compounds are very sensitive to moisture and must be handled in an inert atmosphere.*

A mixture of 13.6 g. (0.20 mole) of imidazole and 24.2 g. (0.15 mole) of hexamethyldisilazane (Alfa Inorganics, P.O. Box 159, Beverly, Mass. 01915) containing two drops of concentrated sulfuric acid is boiled under reflux for 2 hours in a 100-ml. round-bottomed flask equipped with a condenser surmounted by a nitrogen flush system. The condenser is replaced by a simple still head. The 1-(trimethylsilyl)imidazole distills at 92° at 12 torr.

2. Hexamethyldisilthiane

$$2(CH_3)_3Si-N \underset{\diagdown}{\overset{=N}{\diagup}} + H_2S \longrightarrow [(CH_3)_3Si]_2S + 2H-N \underset{\diagdown}{\overset{=N}{\diagup}}$$

■ **Caution.** *Trimethylsilylimidazole is very sensitive to hydrolysis and must be handled under an inert atmosphere at all times.*

■ **Caution.** *Hexamethyldisilthiane appears to have an effect on the nervous system. Skin contact with the liquid or breathing of the vapor should be avoided. This compound should be handled only in a good hood. It has a very strong odor reminiscent of mercaptans.* If silthiane vapors are allowed to escape into the air, a flood of gas-leak reports will be made and personnel will evacuate the building because they think that there is a gas leak. As a result, no article which has contacted the sulfide should be allowed out in the open room.

The reaction between 1-(trimethylsilyl)imidazole and hydrogen sulfide may be effected either in a standard high-vacuum system or in a bench-top assembly constructed from vacuum rubber tubing, a 1-l. volume bulb, a mercury-bubbler manometer, and a one-necked round-bottomed flask. The round-bottomed flask is used as the reaction flask and is connected through a glass T tube to both the mercury-bubbler manometer and a three-way stopcock. The three-way stopcock allows access to vacuum, dry nitrogen, or dry H_2S as required.

Using a standard all-glass vacuum system, 22.7 g. (0.162 mole) of freshly distilled 1-(trimethylsilyl)imidazole is placed in a 50-ml. one-necked flask fitted with an O-ring joint which contains a magnetic stirring bar. This operation is carried out in a nitrogen-filled dry-bag. The flask is then connected to a high-vacuum system and evacuated. An approximately stoichiometric amount of H_2S gas (dried by passage over P_2O_5 dispersed on glass beads) is allowed to fill the evacuated apparatus. The reaction flask is cooled to $0°$ and stirred for 12 hours. The 1-(trimethylsilyl)imidazole reacts with the H_2S as evidenced by the drop in pressure of the system. The reaction is accompanied by the precipitation of solid white imidazole. The volatile portion of the reaction mixture is separated by passing it through a series of cold baths at $0°$, $-30°$, and $-196°$. Hexamethyldisilthiane stops in the $-30°$ bath and upon repeated fractionation gives a vapor pressure of 3 to 4 torr at room temperature $(25-27°)$. The reaction mixture may be purified by fractional distillation to a boiling point of $162°$. *Anal.* Calcd. for $C_6H_{18}Si_2S$: C, 40.44; H, 10.11; S, 17.96; mol. wt., 178.5. Found: C, 40.31; H, 10.40; S, 17.70; mol. wt., 176.9 (vapor density).

Because of the intense odor of the product, all glassware used in preparing this material should be soaked in sodium hydroxide solution for a number of days before it is taken out of the hood. A solution of nickel acetate in aqueous ethanol quickly removes the odor.

The sulfide is an excellent solvent for stopcock greases of all types. Thus, all stoppers used in the work should be well secured. Stoppers on storage vessels should be checked from time to time. The storage bottle should be kept in a desiccator for safety.

It is often advisable to change clothes and bathe after preparing this compound.

Properties

Hexamethyldisilthiane is a colorless, stinking liquid which boils at $162°$; n_D^{20}, 1.4590; d_4^{20}, 0.851. The proton n.m.r. spectrum[9] shows a singlet methyl resonance at $\tau 9.67$. Its infrared spectrum contains peaks at 3.3, 3.4, 6.9, 7.1, 8.0, 9.95, 11.85, 12.1, 13.2, and 14.5 μm.[11]

B. HEXAMETHYLDISILTHIANE VIA CHLOROTRIMETHYLSILANE

$$2(CH_3)_3SiCl + H_2S + C_5H_5N \longrightarrow [(CH_3)_3Si]_2S + C_5H_5NHCl$$

Procedure

■ **Caution.** *Hexamethyldisilthiane is hydrolytically sensitive, malodorous, and toxic. Use the precautions listed in Sec. A.*

Chlorotrimethylsilane is purified by distillation in a long, packed column collecting the fraction boiling at $57.°$ Pyridine* is dried over potassium hydroxide pellets and distilled, the fraction boiling at $115°$ being collected. Diethyl ether is dried over sodium metal, and the hydrogen sulfide is dried by passing it over phosphorus(V) oxide.

In a cylindrical reaction vessel 58.0 g. (0.73 mole) of dry pyridine is added to 500 ml. of dry diethyl ether, and dry hydrogen sulfide is condensed into this solution using a condenser packed with solid carbon dioxide and acetone until the solution is saturated with hydrogen sulfide.† Chlorotrimethylsilane (80 g., 0.73 mole) is added slowly over a period of 90 minutes from a bypass dropping funnel while hydrogen sulfide is still being condensed into the solution, as shown in Fig. 8. With occasional swirling of the reaction vessel, hydrogen sulfide is condensed into the solution for a further 10 hours. The mixture is left overnight and is resaturated with hydrogen sulfide for 4 hours. The hydrogen sulfide is then allowed to boil out of solution at room temperature,‡ and the condenser and dropping funnel are replaced by stoppers. The reaction vessel is inverted, and

---

*Triethylamine can be used with success equal to pyridine in both preparations.

†This may be determined when hydrogen sulfide is seen to bubble out of solution.

‡Heating to remove dissolved hydrogen sulfide causes dissociation of pyridine hydrochloride and reverses the reaction. Excess H_2S can be conveniently trapped in ethanolic nickel acetate using an inverted funnel.

the cap at the base of the vessel is replaced by a coarse-porosity frit connected to a three-necked flask. The apparatus is reinverted, and the liquid is allowed to run into the flask, leaving the pyridine hydrochloride on the sinter. This is washed for 8 hours, which involves continuously circulating the solvent by cyclic extraction, as indicated in Fig. 9.

Subsequent distillation of the solution collected in the flask yields ether, unreacted chlorotrimethylsilane, pyridine, and 25.0 g. of hexamethyldisilthiane (38%), b.p. 162°. *Anal.* Calcd. for $C_6H_{18}S_2Si$: C, 40.6; H, 10.2; S, 18.0. Found: C, 40.7; H, 9.7; S, 17.7. Care is required in the final stages of this distillation to separate the pyridine from the product.

If the precipitate is not washed, the final yield is substantially reduced.

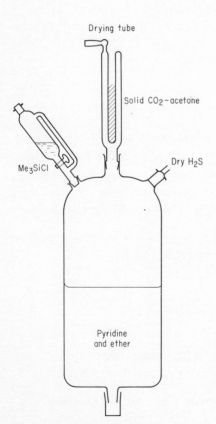

Fig. 8. *Reaction vessel.*

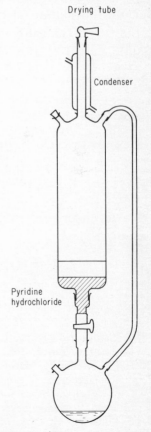

Fig. 9. *Filtration apparatus.*

C. HEXAMETHYLCYCLOTRISILTHIANE

$$3(CH_3)_2SiCl_2 + 3H_2S + 6C_5H_5N \longrightarrow [(CH_3)_2SiS]_3 + 6C_5H_5NHCl$$

Procedure

This reaction is carried out using the same procedure as for hexamethyldisilthiane (Sec. B) with 700 ml. of dry benzene, 174 g. of anhydrous pyridine (2.20 moles), and 129 g. of dichlorodimethylsilane (1.2 moles).

Filtering the reaction mixture in a dry atmosphere is followed by washing with 2 × 250 ml. of dry benzene. The benzene is pumped off with slight warming, and if precipitation occurs in the solution, this is refiltered and the procedure is continued until a clear oil remains. Vacuum distillation gives a crystalline-solid forerun of 2.0 g. of tetramethylcyclodisilthiane, m.p. 110°, and 75 g. of hexamethylcyclotrisilthiane (83%), b.p. 69° at 0.3 torr, m.p. 17°. *Anal.* Calcd. for C_2H_6SSi: C, 26.7; H, 6.7; S, 35.6. Found: C, 26.8; H, 6.6; S, 35.5.

Properties

Hexamethylcyclotrisilthiane is a white, waxy solid which melts at 17° to a colorless liquid, b.p. 60° at 0.3 torr, n_D^{20}, 1.5530, with a singlet peak at $\tau 9.31$ in the proton n.m.r. spectrum. It is exceedingly malodorous and sensitive to water.[9] The cyclotrisilthiane is thermally unstable, undergoing ring contraction at about 200° to tetramethylcyclodisilthiane.

$$2[(CH_3)_2SiS]_3 \longrightarrow 3[(CH_3)_2SiS]_2$$

References

1. C. Eaborn, *J. Chem. Soc.,* **1950**, 3077.
2. C. Eaborn, *Nature,* **165,** 685 (1950).
3. E. Larsson and R. Marin, *Acta Chem. Scand.,* **5,** 964 (1951).
4. G. Champetier, Y. Etienne, and R. Kullmann, *Compt. Rend.,* **234,** 1985 (1952).
5. E. W. Abel, *J. Chem. Soc.,* **1961**, 4933.
6. M. Field, W. Sundermeyer, and O. Glemser, *Chem. Ber.,* **97,** 620 (1964).
7. E. Louis and G. Urry, *Inorg. Chem.,* **7,** 1253 (1968).

8. T. Nomura, M. Yokoi and K. Yamasaki, *Proc. Japan. Acad.,* **29,** 342 (1953); L. S. Moody, U.S. Patent 2,567,742.

9. E. W. Abel and D. A. Armitage, Organosulfur Derivatives of Silicon, Germanium, Tin and Lead, in "Advances in Organometallic Chemistry," Stone and West (eds.), Vol. 5, p. 1, Academic Press, Inc. New York, 1967.

10. L. Birkofer and A. Ritter, *Angew. Chem. Int. Ed.,* **4,** 417 (1965).

11. H. Kriegsman, *Z. Elektrochem.,* **61,** 1088 (1957).

47. POLYATOMIC CATIONS OF SULFUR, SELENIUM, AND TELLURIUM

Submitted by P. A. W. DEAN,* ·R. J. GILLESPIE,* and P. K. UMMAT*
Checked by F. SEEL† and M. R. BUDENZ†

The preparation and properties of homopolyatomic cations of the Group VI elements have been reviewed recently.[1] The general method described below, involving the oxidation of elemental sulfur, selenium, and tellurium with either antimony pentafluoride or arsenic pentafluoride in liquid sulfur dioxide, is convenient for the preparation of compounds containing these polyatomic cations. The procedure is basically that briefly described previously[2] for the preparation of $Se_8(Sb_2F_{11})_2$.

General Procedures

The apparatus used in the preparations is shown in Fig. 10. It consists of two 100-ml. flasks *A* and *B*, open at the top via 3-cm.-long 6-mm.-o.d. glass tubes *C* and *D*, and joined by a tube with a 10-mm. fine- or medium-grade sinter *E* and a constriction *F*. *A* has in it a 25-mm. Teflon-covered magnetic stirrer bar. *C* and *D* can be closed using a Teflon Swagelock fitting with either a 6-mm. glass blank or a 6-mm. tube separated from a standard joint by an intervening stopcock greased with Kel-F grease.

*Department of Chemistry, McMaster University, Hamilton, Ontario, Canada.

†Institut für Anorganische Chemie der Universität des Saarlandes, 66 Saarbrücken, West Germany.

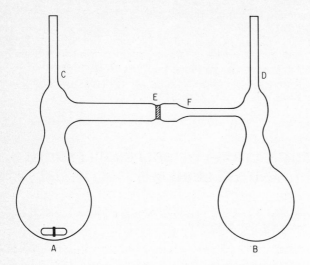

Fig. 10. Apparatus for the preparation and extraction of polychalcogen cation salts in liquid sulfur dioxide.

■ **Caution.** *These reactions are hazardous in several ways and should be carried out only after thorough preparation and with proper protective equipment as given below!*

Selenium and tellurium and their compounds produce very unpleasant and dangerous physiological reactions. They are hazardous both through inhalation as dusts and through skin absorption from solution. Because of these properties and because of the toxicity of the sulfur dioxide used as a solvent, the reactions should be conducted in an efficient fume hood. Rubber gloves and apron are recommended to prevent contact with solutions containing selenium and tellurium compounds as well as the dangerous pentafluorides of antimony and arsenic.

Liquid sulfur dioxide solutions can generate up to 5 atmospheres pressure at room temperature. Therefore, the handling of such solutions in sealed glass equipment requires careful shielding including the use of face shields.

Oxidation Reactions with Antimony Pentafluoride. An appropriate amount of the finely powdered element* is introduced into *A* (Fig.

---

*99.99 % purity, available from K and K Laboratories, 121 Express St., Plainview, N.Y. 11803.

10). The apparatus is closed with the blank at C and the standard-joint stopcock ensemble at D, evacuated, and flamed at those parts where the element will not simultaneously be heated. In a dry-box a weighed* amount of antimony pentafluoride† is added to flask B using an all-glass syringe‡ (because of the viscosity of SbF_5 it is carefully poured into the neck of the syringe and not sucked up in the normal manner). Tubes C and D are closed as before, flask B is cooled in liquid nitrogen, and the whole apparatus is evacuated. About 30 ml. of anhydrous sulfur dioxide (dried over P_2O_5 for several days) is condensed onto the cooled SbF_5, and the 6-mm. exit tubes are heat-sealed. The apparatus is allowed to warm to room temperature when a vigorous exothermic reaction occurs between SbF_5 and the solvent to give $SbF_5 \cdot SO_2$.[3] When this reaction has subsided, some swirling may be necessary to dissolve the SO_2 adduct completely. The SbF_5 solution is poured through the sinter E onto the powered element. Immediate reaction occurs, and except in the case of $Se_4(Sb_2F_{11})_2$, the reaction is completed by stirring at room temperature for 1 hour. To prevent transfer of SO_2 from A to B, the top of the flask A should be wrapped with several turns of rubber tubing carrying cold tap water. Separation of the SO_2-soluble product is achieved by extracting the contents of A with SO_2 until there remains a colorless or nearly colorless solution in A. During this process the solvent is transferred from B to A by cooling A under cold tap water. Finally volatiles are transferred to A by cooling in liquid N_2 for 1 day, while keeping B at *ca.* 60°. Flask B is separated by heat-sealing at F. The products of preparations involving excess SbF_5 are dried to constant weight on the vacuum line.

Oxidation Reactions with Arsenic Pentafluoride. Arsenic penta-fluoride (Ozark Mahoning Company) is handled on a calibrated Pyrex vacuum line equipped with a mercury manometer. The vacuum line is thoroughly flamed out under vacuum before use. The joints on

---

*If weighing in a dry-box is not possible, the SbF_5 may be added by volume; 1 ml. = 3 g. is a good approximation.

†The commercial material (Ozark Mahoning Company) is distilled twice in a thoroughly dried glass apparatus under an atmosphere of dry air, and stored in Teflon bottles.

‡A suitable syringe may be made by drawing out the glass stub of a 5-ml. syringe into a needle *ca.* 5 cm. long and 0.75 mm. i.d.

the line are lubricated with Kel-F grease because AsF_5 attacks ordinary Apiezon grease. Arsenic pentafluoride attacks mercury rapidly but becomes protected by a layer of product. The technique of carrying out a reaction with arsenic pentafluoride is essentially the same as described above for antimony pentafluoride. The reactions with AsF_5 are more convenient than with SbF_5 for three reasons: (1) It is not necessary to use a dry-box, because all handling of AsF_5 may be carried out on the vacuum line. (2) The reduced product of arsenic pentafluoride, arsenic trifluoride, AsF_3, is volatile (b.p. 57–58°) and therefore easily removed under vacuum. (3) Arsenic pentafluoride normally reacts to give a simple anion $[AsF_6]^-$, which is conveniently characterized by Raman, infrared, or ^{19}F n.m.r. spectroscopy.[4] The residual material from either type of oxidation can be disposed of by freezing A in liquid N_2 and breaking open at C in a fume hood so that volatiles may escape on warming. Unreacted selenium and tellurium are dissolved in aqua regia and kept in a separate residues bottle. (■ **Caution.** *Avoid contact of the acid solution with the skin.*) The $SbF_5 \cdot SO_2$ adduct can be decomposed by very careful addition of water.

A. OCTASULFUR μ-FLUOROBIS(PENTAFLUORO-ANTIMONATE(V)), $S_8(Sb_2F_{11})_2$, AND OCTASULFUR HEXAFLUOROARSENATE(V), $S_8(AsF_6)_2$

$$S_8 + 5SbF_5 \longrightarrow S_8(Sb_2F_{11})_2 + SbF_3$$

$$S_8 + 3AsF_5 \longrightarrow S_8(AsF_6)_2 + AsF_3$$

Procedure $S_8(Sb_2F_{11})_2$

Antimony pentafluoride (5.00 g., 0.023 mole) and sulfur (1.00 g., 0.031 mole) are added to flasks B and A, respectively. A sulfur dioxide solution of antimony pentafluoride is prepared as described in the general procedure. When the pentafluoride solution is first added to the sulfur, a red color develops because of the formation of the $(S_{16})^{2+}$,[5] but the solution becomes deep blue as the reaction proceeds. Room-temperature extraction of the SO_2-soluble $S_8(Sb_2F_{11})_2$ is achieved as described above (three extractions leave a white product in A which is SbF_3 containing a small amount of $SbF_3 \cdot SbF_5$), followed by removal of the volatiles into A. The neck of the separated flask B is opened in the dry-box by cracking at the constriction, and the product is removed. The powdered product is pumped to constant weight. The yield is typically 0.0035 mole (90% based on sulfur). *Anal.* Calcd. for $S_8Sb_4F_{22}$: S, 22.03; Sb, 41.60; F, 36.00. Found: S, 21.21; Sb, 40.71; F, 37.74.

Procedure $S_8(AsF_6)_2$

Arsenic pentafluoride (2.08 g., 0.012 mole) is condensed onto powdered sulfur (1.047 g., 0.033 mole) and frozen sulfur dioxide at $-196°$, and the reaction vessel is heat-sealed at C and D. The mixture is allowed to warm to room temperature. The solution immediately develops a red color owing to $(S_{16})^{2+}$, just as in the case of the antimony pentafluoride reaction, but on stirring for 1 hour a blue solution is obtained. The reaction is complete within 3 hours. The SO_2-soluble $S_8(AsF_6)_2$ is extracted until the solution in A is colorless and there is no residue (five extractions are normally sufficient). As described in the general procedure, flask B is heat-sealed at F while keeping A at liquid N_2 temperature. The powdered product obtained from B is further dried to give a yield of 90% or better. In the typical experiment described 2.48 g. (0.0039 mole) was obtained: a 96% yield based on sulfur. *Anal.* Calcd. for $S_8As_2F_{12}$: S, 40.43; As, 23.62; F, 35.94. Found: S, 40.53; As, 23.63; F, 36.12.

Properties

The compounds $S_8(Sb_2F_{11})_2$ and $S_8(AsF_6)_2$ are both shiny blue crystalline solids which rapidly disproportionate in moist air to give elemental sulfur. The former has a sharp m.p. (in a sealed tube) at 128–130° to give a blue melt, whereas the latter shows signs of decomposition around 105°, but at 130° the whole mass melts to give a green-blue color. Freshly prepared solutions of both compounds in HSO_3F show the characteristic band of S_8^{2+} at 585 nm.[5] The crystal structure of $S_8(AsF_6)_2$ has been recently determined by Davies *et al.*[6] by x-ray diffraction. The $(S_8)^{2+}$ ion is a ring with an endo-exo conformation and C_s symmetry.

B. OCTASELENIUM μ-FLUOROBIS(PENTAFLUORO-ANTIMONATE(V)), $Se_8(Sb_2F_{11})_2$, AND OCTASELENIUM HEXAFLUOROARSENATE(V), $Se_8(AsF_6)_2$

$$Se_8 + 5SbF_5 \longrightarrow Se_8(Sb_2F_{11})_2 + SbF_3$$

$$Se_8 + 3AsF_5 \longrightarrow Se_8(AsF_6)_2 + AsF_3$$

Procedure $Se_8(Sb_2F_{11})_2$

The starting materials are 2.00 g. (0.025 mole) of selenium and 3.00 g. (0.014 mole) of antimony pentafluoride. Following the reaction as described in the section on General Procedures, about 12 room-temperature extractions are necessary before a clear solution is left in A. In a typical experiment using the above amounts of reactants, 3.84 g. (0.0025 mole) of deep-green $Se_8(Sb_2F_{11})_2$ was obtained (90% based on SbF_5). *Anal.* Calcd. for $Se_8Sb_4F_{22}$: Se, 41.4; Sb, 31.6; F, 27.3. Found: Se, 40.4; Sb, 30.5; F, 28.7.

Procedure $Se_8(AsF_6)_2$

In a typical reaction, 0.78 g. (0.0046 mole) of arsenic pentafluoride (■ **Caution.**) is condensed onto the 0.97 g. (0.01 mole) of powdered selenium in SO_2 at $-196°$. Following the reaction, as detailed for $S_8(AsF_6)_2$, about five room-temperature extractions are necessary. In a typical experiment using the above amounts of reactants 1.54 g. (0.0015 mole) of semicrystalline $Se_8(AsF_6)_2$ was obtained (96% yield based on selenium). *Anal.* Calcd. for $Se_8As_2F_{12}$: Se, 62.6; AsF_6, 37.4. Found: Se, 63.1; AsF_6, 37.0 (by precipitation as $[Ph_4As][AsF_6]$).

Properties

$Se_8(Sb_2F_{11})_2$ and $Se_8(AsF_6)_2$ are deep-green, semicrystalline materials. Both decompose in a sealed m.p. tube to red elemental selenium—the decomposition temperatures are $193°$ and $167°$, respectively. Both compounds dissolve in HSO_3F to give a green solution which has the characteristic absorption bands of $(Se_8)^{2+}$ at 295, 470, and 685 nm.[2] The structure of $(Se_8)^{2+}$ is similar to that described for $(S_8)^{2+}$ above.[7]

C. TETRASELENIUM μ-FLUOROBIS(PENTAFLUORO-ANTIMONATE(V)), $Se_4(Sb_2F_{11})_2$ AND TETRASELENIUM HEXAFLUOROARSENATE(V), $Se_4(AsF_6)_2$

$$Se_8 + 10SbF_5 \longrightarrow 2Se_4(Sb_2F_{11})_2 + 2SbF_3$$

$$Se_8 + 6AsF_5 \longrightarrow 2Se_4(AsF_6)_2 + 2AsF_3$$

Procedure $Se_4(Sb_2F_{11})_2$

Starting with 1.00 g. (0.013 mole) of selenium and 5.00 g. (0.023 mole) of antimony pentafluoride, the procedure is at first identical with that for $Se_8(Sb_2F_{11})_2$. However, when the room-temperature reaction is complete, the whole apparatus is placed in a 60° oven for 12 hours. A bright yellow solid and a yellow solution are obtained. As a safety precaution the oven is switched off and the apparatus is allowed to cool to room temperature before being removed. Three room-temperature extractions transfer all the $Se_4(Sb_2F_{11})_2$ into flask *B*. After sulfur dioxide has been distilled, the product is removed from *B* and pumped to constant weight. The yield is quantitative. *Anal.* Calcd. for $Se_4Sb_4F_{22}$: Se, 25.87; Sb, 39.89; F, 34.24. Found: Se, 27.04; Sb, 38.54; F, 33.9.

Procedure $Se_4(AsF_6)_2$

In a typical reaction, 1.73 g. (0.01 mole) of arsenic pentafluoride (■ **Caution.**) is condensed onto 1.07 g (0.0136 mole) of powdered selenium in liquid SO_2. The apparatus is heated gradually to 60° for 6 days, during which time a yellow compound precipitates from a green solution. The apparatus is kept at this temperature until a colorless solution and a bright yellow precipitate is obtained. After removing the volatiles, the product is removed from *A* and pumped to constant weight. The yield is quantitative. *Anal.* Calcd. for Se_4-$(AsF_6)_2$: Se, 45.53; $AsF_6{}^-$, 54.47. Found: Se, 44.95; $AsF_6{}^-$, 55.30.

Properties

Both $Se_4(Sb_2F_{11})_2$ and $Se_4(AsF_6)_2$ readily dissolve in HSO_3F to give a yellow solution which has a characteristic absorption at 410 nm. $Se_4(Sb_2F_{11})_2$ melts in the range 154.5–159.5° to give a yellow liquid, whereas $Se_4(AsF_6)_2$ decomposes around 115° to give a green solid. The structure of $(Se_4)^{2+}$ has been shown to be square planar, both by single-crystal x-ray crystallography and by vibrational spectroscopy.[8,9] The $(Se_4)^{2+}$ cation has a characteristic Raman line at *ca.* 327 cm.$^{-1}$ and *ca.* 188 cm.$^{-1}$.

D. TETRATELLURIUM μ-FLUOROBIS(PENTAFLUORO-ANTIMONATE(V)), $Te_4(Sb_2F_{11})_2$, AND TETRATELLURIUM HEXAFLUOROARSENATE(V), $Te_4(AsF_6)_2$

$$4Te + 5SbF_5 \longrightarrow Te_4(Sb_2F_{11})_2 + SbF_3$$

$$4Te + 3AsF_5 \longrightarrow Te_4(AsF_6)_2 + AsF_3$$

Procedure $Te_4(Sb_2F_{11})_2$

A mixture of 2.81 g. (22 mmoles) of tellurium and 10.25 g. (47 mmoles) of antimony pentafluoride in liquid sulfur dioxide is stirred for 3 days at −23°. The solution develops a red color after about 2 hours of stirring at −23°. The solution is filtered to leave a yellow residue, which is repeatedly washed with sulfur dioxide until the washings are colorless and all the red compound has been removed. In this particular reaction, since both red and yellow products are needed, flask *B* is not heat-sealed but is closed with a Nupro* Teflon valve. After all the SO_2 has been removed under vacuum, flask *B* is heat-sealed at *F*. The yellow product is discussed below. The product obtained from flask *B* is a semicrystalline red material; the yield is 6.0 g. (4.2 mmoles); 75% yield based on tellurium. *Anal.* Calcd. for $Te_4Sb_4F_{22}$: Te, 35.97; Sb, 34.45; F, 29.60. Found: Te, 35.88; Sb, 34.23; F, 32.21.

Procedure $Te_4(AsF_6)_2$

Arsenic pentafluoride (0.66 g., 3.9 mmoles) is condensed onto powdered tellurium (0.66 g., 5.5 mmoles) and frozen sulfur dioxide at −196°, and the mixture is allowed to warm up to room temperature. A deep-red solution containing the red solid is obtained after stirring

*Supplied by Nupro Company, Cleveland, Ohio.

for 1 hour. The stirring is continued for a further 24 hours at room temperature to complete the reaction. The solid obtained from flask *B* is a crystalline deep-red material. The yield is quantitative. *Anal.* Calcd. for $Te_4As_2F_{12}$: Te, 57.46. Found: Te, 57.64.

Properties

The absorption spectra of freshly prepared solution of Te_4-$(Sb_2F_{11})_2$ and $Te_4(AsF_6)_2$ in 100% HSO_3F show the characteristic peak of $(Te_4)^{2+}$ at 510 nm. The Raman spectrum gave two peaks at 219 and 139 cm.$^{-1}$ which have been assigned to the A_{1g} and B_{2g} modes of $(Te_4)^{2+}$ cation. The compound $Te_4(Sb_2F_{11})_2$ melts to a red liquid around 114°; $Te_4(AsF_6)_2$ appears to decompose at approximately 100°.

E. $Te_n(SbF_6)_n$

The yellow product insoluble in sulfur dioxide at $-23°$ that is obtained in the preparation of $Te_4(Sb_2F_{11})_2$ is heated under vacuum at 100° for 36 hours to remove antimony trifluoride. The yield of the residual yellow compound is 2 g. (5.5 mmoles, 25% yield based on tellurium). *Anal.* Calcd. for $TeSbF_6$: Te, 35.11; Sb, 33.50; F, 31.17. Found: Te, 35.42; Sb, 33.45; F, 32.13.

Properties

The yellow compound, $Te_n(SbF_6)_n$, is not stable in HSO_3F at room temperature since it gives a red solution containing the $(Te_4)^{2+}$ cation which is presumably formed by a disproportionation reaction. However, a stable yellow solution is obtained in HSO_3F/SbF_5 which has a higher acidity than HSO_3F. The absorption spectrum in this medium shows a strong band at 250 nm. and weak bands at 350 and 420 nm. The Raman spectrum of $Te[SbF_6]$ and its solution in HSO_3F at low temperature have a characteristic line at 699 cm.$^{-1}$. The compound, $Te[SbF_6]$, decomposes to black tellurium around 120°. Although it has been clearly established that this compound contains tellurium in the $+1$ oxidation state, it has not yet been possible to establish the value to *n* with certainty, although there is some evidence that *n* is not less than 4.

General Properties

Because these cations are strong Lewis acids, they are stable only under very weakly basic conditions and can be isolated only as the salts of very weakly basic anions such as SbF_6^-, $Sb_2F_{11}^-$, or AsF_6^-. The cations instantaneously disproportionate in basic media such as H_2O. It is therefore essential to handle the compounds of these cations under rigorously anhydrous conditions, i.e., in a very good dry-box. The cations are, however, stable in acidic media such as HSO_3F, oleum, and HSO_3F/SbF_5 in which the most basic species is the solvent anion. However, $(S_8)^{2+}$ and $(Te_n)^{n+}$ are not stable in HSO_3F.

References

1. R. J. Gillespie and J. Passmore, *Accounts Chem. Res.*, **4**, 413 (1971).
2. R. J. Gillespie and P. K. Ummat, *Can. J. Chem.*, **48**, 1239 (1970).
3. E. E. Aynsley, R. D. Peacock, and P. L. Robinson, *Chem. Ind.*, **1951**, 1117.
4. C. G. Davies, P. A. W. Dean, R. J. Gillespie, and P. K. Ummat, *Chem. Commun.*, **1971**, 782.
5. R. J. Gillespie and J. Passmore, *Chem. Commun.*, **1969**, 1333.
6. C. G. Davies, R. J. Gillespie, J. J. Park, and J. Passmore, *Inorg. Chem.*, **10**, 2781 (1971).
7. R. K. McMullan, D. J. Prince, and J. D. Corbett, *Chem. Commun.*, **1969**, 1438.
8. I. D. Brown, D. B. Crump, R. J. Gillespie, and D. Santry, *ibid.*, **1968**, 853.
9. R. J. Gillespie and G. P. Pez, *Inorg. Chem.*, **8**, 1229 (1969).

48. TETRAETHYLAMMONIUM TRICHLOROGERMANATE-(1−) AND TRICHLOROSTANNATE(1−)

Submitted by G. W. PARSHALL*
Checked by W. L. JOLLY,† M. IANNONE,† and P. KOO†

Complexes of the $[SnCl_3]^-$ anion and, to a lesser extent, the $[GeCl_3]^-$ anion have found extensive application in homogeneous catalysis.[1] All the Group VIII metals have been reported to form

*Central Research Department, E. I. du Pont de Nemours & Company, Wilmington, Del. 19898.
†Department of Chemistry, University of California, Berkeley, Calif. 94720.

discrete $[SnCl_3]^-$ complexes.[2] Although many of these derivatives can be prepared from $SnCl_2$ or $GeCl_2$, it is often advantageous to use salts of the tricoordinate anions as starting materials. These salts are conveniently prepared[3-7] by adding a solution of a bulky cation to a hydrochloric acid solution of $SnCl_2$ or $GeCl_2$ as described below. The desired salt precipitates immediately and in good purity because, at the appropriate HCl concentration, nearly all the M^{2+} species are present as $[MCl_3]^-$ anion.[5]

Commercial $SnCl_2 \cdot 2H_2O$ serves quite well in these reactions. Commercially available $GeCl_2$ can also be used analogously, but because of cost and quality problems, it is often desirable to generate $GeCl_2$ *in situ*[6] as illustrated here.

This general method works well for a variety of tetraalkyl and tetraarylammonium, phosphonium, and arsonium salts and can also be used to isolate the $[SnBr_3]^-$ salts from hydrobromic acid solution.

A. TETRAETHYLAMMONIUM TRICHLOROGERMANATE(1−)

$$GeCl_4 \xrightarrow{\text{H}_3\text{PO}_2} GeCl_2 \xrightarrow{[(C_2H_5)_4N]Cl} [(C_2H_5)_4N][GeCl_3]$$

Procedure

■ **Caution.** *The product is moderately air-sensitive. Solutions should be protected from oxygen during both preparation and recrystallization.*

A 500-ml. round-bottomed flask fitted with stirrer, addition funnel, thermometer, and condenser is flushed with nitrogen and charged with a solution of 70 g. of 50% hypophosphorous acid (phosphinic acid) in 80 ml. of $3M$ hydrochloric acid. Germanium(IV) chloride (tetrachlorogermane) (15.1 ml., 0.13 mole) is added over a period of about 15 minutes while the solution is stirred and maintained at 50–60°. (A steam bath is convenient.) The solution is stirred at 85° for 5 hours and then is cooled to about 50°. A solution of 25 g. of tetraethylammonium chloride hydrate in 15 ml. of $3M$ hydrochloric acid is added, and the mixture is cooled to 0°. The product precipitates, sometimes as an oil, sometimes as white needles. The crude product, after decantation of the supernatant liquid, is dissolved in a minimum volume (*ca.* 120 ml.) of absolute ethanol at room temperature. Precipitation by slow addition of 700 ml. of diethyl ether gives a white, crystalline precipitate of 19–21 g. (*ca.* 50%) of tetraethyl-ammonium trichlorogermanate(1 −) which, after an ether wash and drying at 25° at 1 torr, is pure enough for most synthetic applications. Careful recrystallization from 60–75 ml. of warm ethanol gives white needles, m.p. 68.5–70°. *Anal.* Calcd. for $C_8H_{20}Cl_3GeN$: C, 31.07; H, 6.52; N, 4.53. Found: C, 30.82; H, 6.77; N, 4.34.

B. TETRAETHYLAMMONIUM TRICHLOROSTANNATE(1−)[4]

$$SnCl_2 \cdot 2H_2O + (C_2H_5)_4NCl \cdot H_2O \xrightarrow[H_2O]{HCl} (C_2H_5)_4N[SnCl_3]$$

Procedure

A warm, clear (filter, if necessary) solution of 22.6 g. (0.1 mole) of tin(II) chloride dihydrate in 25 ml. of 0.5M hydrochloric acid is stirred magnetically in a 100-ml. suction flask with a slow stream of nitrogen through the side arm to exclude air. A solution of 18.4 g. (0.1 mole) of tetraethylammonium chloride monohydrate in a mini-mum volume (*ca.* 10 ml.) of 0.5M hydrochloric acid is added rapidly. The mixture containing precipitated $[(C_2H_5)_4N][SnCl_3]$ is cooled to 0° and filtered. The crystalline product is dried at 25° at 1 torr and is recrystallized from a small volume of hot absolute ethanol *with thorough exclusion of oxygen*. Ether is added to the warm ethanol solution until *faint* cloudiness occurs. Cooling gives 23.0 g. (63%) of large white crystals of tetraethylammonium trichlorostannate(1 −), m.p. 78–78.5°. *Anal.*[7] Calcd. for $C_8H_{20}Cl_3NSn$: Cl, 29.99; Sn, 33.39. Found: Cl, 30.82; Sn, 32.74.

This general procedure also works well for preparation of the tetramethylammonium salt[7] (m.p. 346°, decomposes), the tetra-butylammonium salt (m.p. 59–60°), the trimethylphenylammonium salt (m.p. 106–106.5°), and the methyltriphenylphosphonium salt[7] (m.p. 104–106°). The reaction of methyltriphenylphosphonium bromide with $SnBr_2$ in aqueous hydrobromic acid gives $[(C_6H_5)_3-PCH_3][SnBr_3]$ as off-white crystals, m.p. 113–114°.

Properties

Tetraethylammonium trichlorostannate(1 −) is moderately stable to air in the solid state, but oxygen should be rigorously excluded for long-term storage and from solutions. It melts at 78–78.5° without decomposition to a colorless liquid, d^{100} 1.48 g./ml. Infrared, Raman, and Mössbauer spectral studies have been reported[8] for $[(C_2H_5)_4N]$-$[SnX_3]$ and $[SnX_2Y]$ salts (X, Y = Cl, Br, I).

References

1. R. D. Cramer, E. L. Jenner, R. V. Lindsey, Jr., and U. G. Stolberg, *J. Am. Chem. Soc.*, **85**, 1691 (1963); J. C. Bailar and H. Itatani, *ibid.*, **89**, 1592 (1967); G. W. Parshall, U.S. Patent 3,565,823 (1971).
2. J. G. Young, *Advan. Inorg. Chem. Radiochem.*, **11**, 91 (1968).
3. M. P. Johnson, D. F. Shriver, and S. A. Shriver, *J. Am. Chem. Soc.*, **88**, 1588 (1966).
4. F. N. Jones, *J. Org. Chem.*, **32**, 1667 (1967).
5. G. P. Haight, J. Zoltewicz, and W. Evans, *Acta Chem. Scand.*, **16**, 311 (1962).
6. P. S. Poskozim, *J. Organometallic Chem.*, **12**, 115 (1968).
7. R. V. Lindsey, Jr., and U. G. Stolberg, unpublished work.
8. R. J. H. Clark, L. Maresca, and P. J. Smith, *J. Chem. Soc. (A)*, **1970**, 2687.

49. WEAK AND UNSTABLE ANIONIC BROMO AND IODO COMPLEXES

Submitted by JACK L. RYAN*
Checked by W. C. WOLSEY† and T. MOELLER‡

The complexes of lanthanide ions with the heavy halides, particularly iodide, are often extremely weak. In many cases preparation of these complexes in or from solution is very difficult if not impossible. The donor properties of any otherwise satisfactory solvent or the donor properties of difficult to remove solvent impurities are such that they effectively compete with the halide ion and prevent its entering the metal coordination sphere. The preparative method discussed here, which is applicable to the preparation of salts of bromo

*Battelle, Pacific Northwest Laboratories, Richland, Wash. 99352. This paper is based on work performed under Contract No. AT(45-1)-1830 for the U.S. Atomic Energy Commission.
†Macalester College, St. Paul, Minn. 55105.
‡Arizona State University, Tempe, Ariz. 85281.

or iodo complexes with large organic cations, completely avoids the use of solvents which can coordinate to the metal ions.

Complexes of relatively strongly oxidizing metal ions with the more reducing halide ions are not prepared easily because the halide ion is oxidized by the metal ion. The low-temperature ($< 25°$) method discussed here allows the preparation of bromo and iodo complexes of oxidizing metal ions which could not be prepared by other means. The complex is formed directly as a solid salt in which crystal-lattice energy gives stability.

The method described here involves the reaction described by the equation:

$$R_mMCl_n + nHX \rightleftharpoons R_mMX_n + nHCl$$

in which X is Br or I. Experimentally it is observed[1-4] that this equilibrium is established rapidly even at liquid HX temperatures if R is a large organic cation. The reaction is driven well to the right even when M is a lanthanide ion because of the less negative free energies of formation of HBr and HI than of HCl and because of the greater volatility of HCl than HBr or HI. This reaction can also be carried out in solvents such as acetonitrile and nitromethane, and addition of excess bromide plus HBr to a solution of $[MCl_n]^{m-}$ converts it completely to $[MBr_n]^{m-}$. Hydrogen iodide reacts rapidly with the solvents at $25°$, but the conversion of chloro complexes in solution to iodo complexes can be carried out near the freezing point of the solvent.

Although in theory it should be possible to prepare almost all anionic bromo and iodo complexes by this method, the real value of the method lies in the preparation of those complexes that cannot be made by other methods or can be made only with difficulty. The increased reactivity, due to the very fine crystal size of the products obtained by this method, probably makes it desirable that bromo and iodo complexes which are easily prepared from solutions be made in that manner. Typical examples of bromo and iodo complexes which can be prepared by this method are discussed here.

■ **Caution.** *All the preparations including those of the chloro complexes should be carried out in an efficient hood.*

A. STARTING MATERIALS—CHLORO COMPLEXES

$$CeO_2 \cdot xH_2O + 4H^+ + nCl^- \longrightarrow [CeCl_n]^{4-n} + (2 + x)H_2O$$

$$[CeCl_n]^{4-n} + 2[(C_6H_5)_3PH]^+ + (6 - n)Cl^- \longrightarrow [(C_6H_5)_3PH]_2[CeCl_6]$$

Procedure

1. Triphenylphosphonium Hexachlorocerate(2 –)

Ammonium hexanitratocerate(2 –) (2 g.) is dissolved in about 30 ml. of water, and the solution is neutralized with excess concentrated NH_4OH. The mixture is centrifuged and the solid is washed with water until the wash is neutral to pH paper. This wet, freshly prepared, and washed cerium(IV) oxide hydrate is cooled to $0°$ in an ice bath, and 30 ml. of $0°$ or colder (preferably as low as $- 10°$) HCl-saturated ethanol (absolute ethanol saturated with HCl gas at $25°$) is added with cooling. The freshly prepared hydrated oxide dissolves immediately to form an orange-red solution. To the cold solution is added immediately a solution of triphenylphosphonium chloride prepared by adding an excess (2.5 g.) of triphenylphosphine* to 15 ml. of absolute ethanol and saturating with HCl gas at $25°$. This solution should be prepared before the cerium(IV) solution and should be no warmer than $0°$ when it is added to the cerium(IV) solution. The bright yellow product which precipitates immediately is isolated by filtration of the cold solution on a medium-porosity frit. It is washed with cold ($0°$ or lower) HCl-saturated or almost saturated ethanol, in which it is moderately soluble, followed by a small amount of cold absolute ethanol. The compound is dried as quickly as possible under vacuum. Since cerium(IV) can oxidize HCl, all solutions are kept cold and all steps of the preparation in which Ce(IV) is in contact with HCl are carried out as quickly as possible for this reason. *Anal.* Calcd. for $C_{36}H_{32}CeCl_6P_2$: Ce, 15.9. Found: Ce, 16.3. The salts $[(C_6H_5)_4As]_2[CeCl_6]$ and $[(C_2H_5)_4N]_2[CeCl_6]$ are also prepared in the same manner except that $12M$ hydrochloric acid is used as a solvent in the preparation and washing of the latter. The solid subsequently is washed with acetone.

*Available from Alfa Inorganics, P.O. Box 159, Beverly, Mass. 01915.

2. Triphenylphosphonium Hexachlorolanthanate(3 −) Salts[6]

$$Ln_2O_3 + 6H^+ + 2nCl^- \longrightarrow 2[LnCl_n]^{3-n} + 3H_2O$$

or
$$LnOCl + 2H^+ + nCl^- \longrightarrow [LnCl_n]^{3-n} + H_2O$$

$$[LnCl_n]^{3-n} + 3[(C_6H_5)_3PH]^+ + (6-n)Cl^- \longrightarrow [(C_6H_5)_3PH]_3[LnCl_6]$$

For the light lanthanides (through Eu except for Pr) the oxide (about 0.75 to 1.5 g.) is placed in 25 ml. of absolute ethanol and HCl gas is added until the hot solution is saturated. The heating which results from HCl addition is usually sufficient to dissolve the oxide, but if not, the solution should be heated until the oxide dissolves. (In the case of praseodymium, Pr_6O_{11} is dissolved in boiling concentrated aqueous HCl, the solution is taken to dryness, and the residue is heated strongly on a hot plate. The residue is dissolved in hot HCl-ethanol as with the oxides of the other light lanthanides.) The hot, HCl-saturated ethanol solutions are cooled to 25° or lower and again saturated with HCl gas. It is not necessary to reach complete equilibrium HCl saturation at 25°, but sufficient HCl should be added so that the solution is saturated at about 40–60°. To the lanthanide solutions in HCl-ethanol is added HCl-ethanol containing 125 % of the stoichiometric amount of triphenylphosphine hydrochloride. Such a solution is prepared by adding HCl gas to triphenylphosphine in 25 ml. of absolute ethanol to the point of HCl saturation at not greater than about 60°.

With the heavy lanthanides, the oxides do not dissolve readily in HCl-ethanol. The oxides (2.5 g.) are dissolved in boiling, concentrated aqueous HCl in a small beaker. With some oxides, even dissolution in aqueous HCl is fairly slow and evaporation to a concentrated lanthanide chloride solution of higher boiling point speeds the dissolution of the last traces of oxide. The aqueous solutions are taken to dryness, and the residues are heated strongly ($\sim 300°$) to effect complete dehydration and conversion to LnOCl and $LnCl_3$. The residue is transferred as completely as possible to a suitable container, 15 ml. of absolute ethanol is added, and the solids are dissolved by addition of HCl gas with further heating if necessary. The solutions are cooled and HCl gas is added until complete saturation at 25° is achieved. Ten milliliters of absolute ethanol is added to produce a final solution of approximately 25 ml. volume and 60 % HCl saturation. A solution of triphenylphosphonium chloride is prepared by adding 15 ml. of absolute ethanol to 125 % of the stoichiometric amount of triphenylphosphine and completely saturating with HCl gas to its solubility at 25° followed by the addition of 10 ml. of absolute ethanol to give a solution of somewhat over 25 ml. volume and about 60 % of equilibrium HCl saturation at 25°. This solution is then added to the lanthanide solution.

Although a small amount of water can be tolerated, and indeed is produced by dissolution of the oxides and oxychlorides, the solutions should be kept as dry as possible, and exposure to atmospheric moisture should be minimized. One hundred-milliliter screw-capped bottles with polyethylene-lined caps are convenient containers for the preparation because they are easily closed when HCl or reagent addition is not being made. They can be heated in a water bath if necessary to promote dissolution of the Ln_2O_3 or LnOCl *while the caps are quite loose or while fitted with a rubber stopper with a small-diameter vent tube to prevent pressurization.*

The $[(C_6H_5)_3PH]_3[LnCl_6]$ solutions supersaturate badly, and solutions so concentrated that precipitation would result in virtual solidification of the solution may set for days without spontaneous seeding. The lighter lanthanide salts usually seed spontaneously if very concentrated but may be seeded by bubbling HCl through the solution. Once seeded, the solutions should be shaken vigorously to prevent development of dendritic clumps. The light lanthanide salts crystallize fairly rapidly from either hot or cold solutions. With the heavy lanthanides, the solutions at about 25° are seeded with a small crystal of one of the light lanthanide salts crushed on the end of a stirring rod and the solution is shaken vigorously, during which time crystallization proceeds slowly. The yield is improved somewhat by cooling to 0°.

The salts are filtered with suction on a coarse-fritted funnel and washed, in the case of the light lanthanide salts, with HCl-saturated or nearly saturated ethanol and, in the case of the heavy lanthanide salts, with absolute ethanol containing 60 % of the 25° equilibrium saturation concentration of HCl. The complexes dissociate in contact with solvents such as ethanol or acetone not containing HCl. They are also moisture-sensitive, and excess air should not be pulled through the salts. A convenient way of removing excess HCl-ethanol from the rather bulky salts is with a rubber dam, a thin sheet of rubber latex held over and around the top of the filter. The rubber is sucked down on top of the product and presses out excess liquid. The product is dried under a heat lamp with care not to overheat and thus decompose the organic cation. The dry products are transferred to a dry, closable container while still hot and stored over a good desiccant such as $Mg(ClO_4)_2$. Spectral studies often indicate very slight hydrolysis even after prolonged storage (several months) over $Mg(ClO_4)_2$. The yield varies with the lanthanide in question but in some cases is over 90%.

TABLE I.   Analysis of $[(C_6H_5)_3PH]_3[MCl_6]$ Salts

Metal	Calcd. for $C_{54}H_{48}Cl_6MP_3$			Found		
	C	H	M	C	H	M
Pr	56.7	4.24	—	55.8	4.74	—
Nd	56.5	4.23	12.6	55.8	4.25	12.6
Sm	—	—	13.0	—	—	13.0
Dy	55.7	4.16	13.9	57.0	4.3	13.8
Er	55.4	4.15	14.3	55.6	4.35	14.1
Yb	—	—	14.7	—	—	14.6

The light lanthanide salts can be recrystallized, if desired, by dissolving in hot absolute ethanol and saturating with HCl gas. Such recrystallization with the heavy lanthanide salts as well as preparation from solutions deficient in $[(C_6H_5)_3PH]Cl$ (and probably also low in HCl concentration) sometimes results in the salt $[(C_6H_5)_3PH]_2$-$[LnCl_5]$ (possibly solvated before drying). With the heavy lanthanides, a third compound of lanthanide-ion content corresponding to $[(C_6H_5)_3PH]_4[LnCl_7]$ is obtained in lower yield if the preparation is carried out in solutions completely saturated with HCl at 25°. Neither of these stoichiometries has been observed with the light lanthanides.

The pyridinium salts of $[LnCl_6]^{3-}$ can be made by essentially the same method as the triphenylphosphonium salts but are more rapidly hydrolyzed by atmospheric moisture.

3. Tetraethylammonium Tetrachloroferrate(1−) and Tetrachloroaurate(1−)

This salt of the well-known $[FeCl_4]^-$ ion is easily precipitated by adding a small excess of $[(C_2H_5)_4N]Cl$ in $12M$ HCl to a $12M$ HCl solution of iron(III) chloride. It is washed with $12M$ HCl and dried by mild heating. *Anal.* Calcd. for $C_8H_{20}Cl_4FeN$: Fe, 17.0. Found: 17.3. The $[AuCl_4]^-$ salt is prepared similarly from a solution of gold(III) chloride (chloroauric acid) in $12M$ hydrochloric acid.

B. TRIPHENYLPHOSPHONIUM HEXABROMOCERATE(2−)

$$[(C_6H_5)_3PH]_2[CeCl_6] + 6HBr \longrightarrow [(C_6H_5)_3PH]_2[CeBr_6] + 6HCl$$

Procedure

All the salts of $[CeBr_6]^{2-}$ are sensitive to atmospheric moisture and can be handled only very briefly in air. About 1 g. of $[(C_6H_5)_3-PH]_2[CeCl_6]$ is placed in a dry 20-cm. test tube fitted with a three-hole stopper which provides inlets for anhydrous nitrogen* and HBr gas and an exit vent. After being swept thoroughly with the inert gas, the test tube is placed in a Dry Ice bath and anhydrous HBr is condensed in the tube until the Ce(IV) salt is completely immersed. A small continuing flow of inert gas prevents moist air from being drawn back into the system on cooling. ■ **Caution.** *Do not introduce HBr with such force that it flows back into the inert-gas drying tube.* The test tube is removed from the Dry Ice bath and allowed to warm slowly to room temperature while maintaining a very slow sweep of inert gas. A second condensation of HBr onto the sample ensures complete conversion. The test tube, at $25°$, is swept thoroughly with inert gas to remove HBr, and the three-hole stopper is replaced quickly with a solid one. The deep-violet compound can be handled very briefly in laboratory air, but any prolonged handling should be in a dry-bag or dry-box. Yield is essentially 100%, and time required is about one hour or less. Chloride should not be detectable by a standard qualitative method.[5] The salts $[(C_6H_5)_4As]_2-[CeBr_6]$ and $[(C_2H_5)_4N]_2[CeBr_6]$ may be prepared by the same method.

---

*The gas is dried by passing it through a long drying tube filled with $Mg(ClO_4)_2$.

Various salts of the $[UBr_6]^{2-}$ and $[UO_2Br_4]^{2-}$ ions also have been prepared in this manner.[2]

Properties

The $[CeBr_6]^{2-}$ salts gradually decompose in the presence of atmospheric moisture to cerium(III) and $[Br_3]^-$. The triphenylphosphonium salt is more stable than the tetraethylammonium salt and may be stored in a desiccator over $Mg(ClO_4)_2$ for several years without apparent decomposition. The salts can be dissolved in dry nitromethane or acetonitrile containing a small amount of anhydrous HBr to produce deep wine-red solutions of $[CeBr_6]^{2-}$ which decompose at 25° with a half-time of several minutes. However, near the freezing points of the solvents, the half-time is at least several hours.[6]

C. HEXAIODOLANTHANIDE(3 −) SALTS

$$[(C_6H_5)_3PH]_3[LnCl_6] + 6HI \longrightarrow [(C_6H_5)_3PH]_3[LnI_6] + 6HCl$$

Procedure

The preparative procedure is essentially the same as for the bromo complexes discussed in Sec. B, substituting HI for HBr, except that oxygen must be excluded carefully because it reacts with both HI and with the final salts to produce products contaminated with $[I_3]^-$. A simple apparatus for this preparation consists of an 8-in. test tube with three-hole stopper as in Sec. B. The 1/4-in. Monel line from the HI tank is bent over and down and inserted about 4 in. through the stopper. The inert-gas purge line (1/8-in.-o.d. Saran plastic) passes through the stopper and is then inserted back inside the Monel line to the needle valve so that the Monel line can be purged. The 1/8-in.-o.d. Saran vent line is inserted about 1 1/2 in. through the stopper and is long enough to lead to the back of the fume hood. This apparatus minimizes contact of HI with stopcock greases, etc., with which it is quite reactive, but some reaction with the rubber stopper does occur. The stopper and gas lines are inserted through a small slit in the top of a gloved dry-bag, and the bag is taped tightly around the lines just above the stopper so that the product can be handled in dry inert gas without releasing HI into the bag. A Dry Ice–acetone bath, a product container, a spatula, and other items desired for handling the product are placed in the dry-bag. Dry, very-low-oxygen-content inert gas (He is preferred by the author) is used to flush and maintain a continuous slow flow through both the reaction vessel and the dry-bag.

A small amount of dry $[(C_6H_5)_3PH]_3[LnCl_6]$ (where Ln is any of the lanthanides) is placed in the apparatus, and the preparation[2] is carried out in the manner described for bromo complexes in Sec. B. Preparation on a 500-mg. or smaller scale improves the ease with which the last of the excess HI is removed from the product. After thoroughly sweeping out HI, the moisture- and oxygen-sensitive product is handled in dry inert gas in the dry-bag. The yield is essentially 100%, and Cl^- analysis (by x-ray fluorescence) indicates $< 1\%$ residual Cl^-. The pyridinium salts can also be prepared in this manner.

Properties

The $[SmI_6]^{3-}$ salts are bright yellow, $[EuI_6]^{3-}$ salts are deep olive-green, and $[YbI_6]^{3-}$ salts are violet-red because of electron-transfer absorption in the visible region. Other $[LnI_6]^{3-}$ salts are about the same colors as the sesquioxides or the $[LnCl_6]^{3-}$ salts.[6] The salts of the colorless lanthanides generally develop a faint yellow cast when they are warmed to 25° owing to a trace of I_3^-, but they are normally white before being warmed much above the boiling point of HI. The $[LnI_6]^{3-}$ salts are extremely sensitive to moisture and oxygen, with the stability increasing with increasing lanthanide atomic number. They also appear to be somewhat light-sensitive. No solvent could be found in which the $[LnI_6]^{3-}$ complexes were stable toward dissociation even in the presence of excess I^- and/or HI.

D. TETRAETHYLAMMONIUM TETRAIODOFERRATE(1−) AND TETRAIODOAURATE(1−)

Procedure

Both these intensely black salts are prepared[3] from the corresponding tetrachloro salts in the same manner as the hexaiodolanthanide-

(III) salts described in Sec. C. Both Fe(III) and Au(III) are much less strongly A-group in character and form stronger iodo complexes than the lanthanides. As a result, $[(C_2H_5)_4N][FeI_4]$ and $[(C_2H_5)_4N]-[AuI_4]$ are both less sensitive to moisture. They can be handled briefly in laboratory air, and can be made without the use of the gloved bag as for the $[CeBr_6]^{2-}$ complexes in Sec. B, but prolonged exposure, particularly in the case of the iron compound, results in reduction of the metal ion by iodide. Any prolonged handling of the products should be in a gloved bag or dry-box. Qualitative analysis indicates only a trace of residual chloride, and analysis of $[(C_2H_5)_4N][AuI_4]$ gave 24.0% Au vs. a calculated value of 23.6%.

Properties

Tetraethylammonium tetraiodoferrate dissolves in dry acetonitrile, nitromethane, or acetone to produce intense purplish-black solutions of $[FeI_4]^-$. The Fe(III) is reduced in these solutions with a half-time of a few minutes at $25°$, but the rate is decreased markedly by decrease in temperature and the half-time is at least several hours in acetone at $-78°$. Although $[(C_2H_5)_4N][AuI_4]$ appears to be more stable to atmospheric moisture than the Fe salt, the $[AuI_4]^-$ complex decomposes instantly at $25°$ in nonaqueous solvents. At $-78°$, though, $[(C_2H_5)_4N][AuI_4]$ dissolves to some extent in dry acetone or propylene carbonate to produce violet solutions which decompose in a few minutes to produce brown $[I_3]^-$.

E. OTHER IODO COMPLEXES

Procedure

Salts of black $[TiI_6]^{2-}$, white $[ThI_6]^{2-}$, and red $[UI_6]^{2-}$ also have been prepared in the same manner.[2,3] The preparations of $[CuI_4]^{2-}$ and $[CeI_6]^{2-}$ were attempted, but salts of these ions are not sufficiently stable to be observed even at the freezing point of liquid HI. Salts of black $[UI_6]^-$ have been prepared from the chloro salts in this manner[4] but are unstable and decompose rapidly above about $-30°$.

References

1. J. L. Ryan, in "Progress in Coordination Chemistry," M. Cais (ed.), paper D16, p. 220, Elsevier Publishing Company, New York, 1968.
2. J. L. Ryan, *Inorg. Chem.,* **8,** 2053 (1969).
3. J. L. Ryan, *ibid.,* **8,** 2058 (1969).
4. J. L. Ryan, *J. Inorg. Nucl. Chem.,* **33,** 153 (1971).
5. T. R. Hogness, W. C. Johnson, and A. R. Armstrong, "Qualitative Analysis and Chemical Equilibrium," p. 513, Holt, Rinehart and Winston, Inc., New York, 1966.
6. J. L. Ryan and C. K. Jørgensen, *J. Phys. Chem.,* **70,** 2845 (1966).

50. HEXAHALOURANATE SALTS*

Submitted by JACK L. RYAN†
Checked by D. G. DURRETT,‡ H. J. SHERRILL,‡ and J. SELBIN‡

A recent review[1] lists various $[UF_6]^-$ salts which were all prepared by the reduction of UF_6 or from UF_5 which in turn was prepared from UF_6 and UF_4 in a bomb. A very convenient method[2] for the preparation of certain salts of $[UF_6]^-$ is based on the simple replacement of the chloride ions of the $[UCl_6]^-$ complex with fluoride ions using 48% aqueous HF. This method completely avoids the use of highly reactive fluorinating agents.

Salts of the $[UCl_6]^-$ ion have been prepared by precipitation from thionyl chloride solutions of U(V) which were in turn prepared by the prolonged (2 weeks) refluxing of UO_3 with thionyl chloride[3] or by the addition of solid $UCl_5 \cdot TCAC$, where TCAC = trichloroacryloyl chloride.[4] This latter extremely moisture-sensitive compound was in turn prepared by the reaction of U_3O_8 with hexachloropropene.[1,4] The method of preparation described here,[2] which is applicable to the preparation of certain salts of the $[UCl_6]^-$ ion, is based on chlorine oxidation in nitromethane solutions of the easily prepared and stored $[UCl_6]^{2-}$ salts. Some of the known salts of

*This paper is based on work performed under Contract No. AT(45-1)-1830 for the U.S. Atomic Energy Commission.
†Battelle, Pacific Northwest Laboratories, Richland, Wash. 99352.
‡Department of Chemistry, Louisiana State University, Baton Rouge, La. 70803.

$[UCl_6]^-$ cannot be prepared by the exact method described here, but in addition to those described, $[(C_3H_7)_4N][UCl_6]$ has been prepared by this method.[5] The $[UBr_6]^-$ salts are prepared[2] in a similar manner by bromine oxidation of $[UBr_6]^{2-}$ salts in nitromethane. Salts of black $[UI_6]^-$ with large organic cations such as $[(C_6H_5)_4As]^+$ can be prepared by the simple reaction of the corresponding $[UCl_6]^-$ salts with excess anhydrous liquid HI at about -40 to $-50°$, but these salts decompose rapidly above about $-30°$ and have not been isolated from the excess HI.[2] Preparative procedures of $[UI_6]^-$ salts are given elsewhere.[6]

A. HEXAHALOURANATE(2−) SALTS

$$U + 4H^+ + nX^- \longrightarrow [UX_n]^{4-n} + 2H_2$$

$$[UX_n]^{4-n} + 2[R_4N]^+ + (6-n)X^- \longrightarrow [R_4N]_2[UX_6]$$

Procedure

The salts of the U(IV) hexahalo complexes which are used as starting materials for synthesis of $[UX_6]^-$ salts are not commercially available, but are easily prepared[2,7] and can be stored indefinitely. A convenient method of preparing these starts with uranium metal. For the chloro complexes, the metal (pickled first in $15M$ HNO_3 to remove oxide) is dissolved in HCl-saturated acetone, keeping the solution at about 25° or less to minimize acetone degradation (discoloration of the solution is not harmful). A small amount of black residue is easily removed by centrifugation. $[(C_2H_5)_4N]_2[UCl_6]$ is precipitated by adding this solution to an equal volume of solution of $[(C_2H_5)_4N]Cl$ in $12M$ HCl. The precipitate is filtered, washed with $12M$ HCl and then with acetone, and dried under a heat lamp. It is stored over $Mg(ClO_4)_2$. $[(C_2H_5NH)]_2[UCl_6]$ (which is moderately deliquescent) and $[(C_6H_5)_4As]_2[UCl_6]$ are prepared in the same manner except that HCl-saturated ethanol is substituted for $12M$ HCl. This method avoids the difficult-to-centrifuge residues which result from dissolution of U in aqueous or ethanolic HCl. For the bromo complexes, uranium metal is dissolved in concentrated aqueous HBr to produce a moderately concentrated U(IV) solution. The solution is thoroughly centrifuged to remove insoluble residues and is then resaturated with HBr. The U(IV) solution is then added to an equal to several-fold larger volume of HBr-saturated ethanol containing excess $[(C_2H_5)_4N]Br$ or $[(C_6H_5)_4As]Br$ to precipitate $[(C_2H_5)_4N]_2[UBr_6]$ and $[(C_6H_5)_4As]_2[UBr_6]$, respectively. The solubility of the products markedly decreases as the HBr-ethanol to aqueous HBr ratio is increased.* The products are filtered, washed with HBr-ethanol and then acetone-dried, and stored over a good desiccant such as Linde molecular sieve 5A.

B. GENERAL CONSIDERATIONS IN HANDLING $[UX_6]^-$ SALTS

Procedure

All hexahalouranate(1 −) complexes are quite sensitive to moisture, but the sensitivity of the salts to atmospheric moisture does vary appreciably with the cation. Although some of the preparative operations can be carried out in the open if the humidity is not high and if exposure to air is carefully minimized, many of the operations and any prolonged handling of the products should be carried out in a dry-atmosphere gloved bag or in a dry-box. All glassware, solvents, and reagents (with the exception of those used with aqueous HF) *must* be water-free. Any possible source of introduction of liquid water must be avoided. Solvents are dried by contacting with anhydrous $CaSO_4$ or Linde molecular sieve 5A and stored in contact with drying agent. Gases are dried by passing through columns of anhydrous magnesium perchlorate.

C. HEXACHLOROURANATE(1−) SALTS

$$2[UCl_6]^{2-} + Cl_2 \longrightarrow 2[UCl_6]^- + 2Cl^-$$

Procedure

■ **Caution.** *These preparations involve the handling of chlorine solutions under pressure. Because container failure could release the toxic gas violently, the vessels (if glass) should be heated in an efficient fume hood with good shielding.*

*This method of preparation is preferable to that using acetone,[7] since excess $[(C_2H_5)_4N]Br$ can be precipitated by acetone addition and acetone degradation can introduce colored impurities.

1. Tetraethylammonium Hexachlorouranate (1−)

Anhydrous nitromethane (5 ml.) is added to $[(C_2H_5)_4N]_2[UCl_6]$ (2 g.) in a 30- to 50-ml. screw-capped linear polyethylene or glass centrifuge tube or similar sealable container of reasonably strong construction. ■ **Caution.** *Pressure develops in the container because of the decrease in* Cl_2 *solubility on heating. Although the preparation has been carried out by the author in flat-bottomed soft glass bottles without incident, stronger containers are recommended.* The slurry is saturated with dry chlorine at 25° and then capped tightly with a polyethylene-lined cap. The solution is then heated to about 90° for 5 minutes with occasional swirling to redissolve the chlorine, which is boiled out by heating. The solution is cooled to 25° and again saturated with chlorine using a clean, dry chlorine addition tube; then it is sealed and heated to about 90° for 5 minutes. A third addition of chlorine at 25° and heating to 90° guarantees complete oxidation but is usually not necessary. All operations up to this point can be carried out in laboratory air if the humidity is not extremely high and if care is taken to minimize as much as possible the exposure of the solution to the air. The solution (and initial reagents) should not be exposed to the laboratory atmosphere longer than necessary, and contact with liquid water must be avoided. If such precautions are not taken, $[UO_2Cl_4]^{2-}$ contamination of the product may result.

The solution along with a small Dewar of Dry Ice–acetone, about 15 ml. of dry nitromethane, a small porcelain spatula, a 15-ml. coarse-porosity fritted-glass funnel, and a container for the product are transferred to a gloved dry-bag in *an efficient fume hood since* Cl_2 *fumes leak from the bag.* The bag should have an intentional small leak toward the back of the hood so that it cannot be overinflated by a continuous slow flow of gas to the bag, and the inlet gas (dry nitrogen) line should be attached inside the bag to about 1 ft. of flexible tubing ending in a one-hole rubber stopper of a size to fit the top of the fritted funnel. After flushing and inflating the bag, the solution is cooled almost to the freezing point of the solvent. The deep-yellow crystalline product is isolated by pressure filtering the cold solution with the dry inert gas. The product is carefully washed three times with dry nitromethane at its freezing point (−29°), forcing the liquid out of the product with the gas stream between each wash. The yield is controlled by the product solubility, which increases with temperature. The volume of each of the washes should be kept as small as possible (1–3 ml.) consistent with satisfactory washing of the funnel and product. Filtration and washing should be done quickly to minimize the warming of the product and funnel between the various steps, and solution temperatures should be kept close to or at the freezing point. The product is dried by continuing to pass the dry inert gas through it. The yield is about 65%. *Anal.* Calcd. for $[(C_2H_5)_4N][UCl_6]$: U, 41.0; Cl, 36.6. Found: U, 40.7; Cl, 35.9. The purity of the product with regard to uranium(IV) can be checked by the absorption spectrum of the solid.[2] Time required for the preparation, not including setting up of the dry-bag and gas-drying tubes or preparation of the starting materials, is about one hour or less.

2. Tetraphenylarsonium Hexachlorouranate (1 −)

Three grams of $[(C_6H_5)_4As]_2[UCl_6]$ and 5 ml. of nitromethane are used, and the procedure is essentially the same as for preparation of $[(C_2H_5)_4N][UCl_6]$. Two chlorine saturations at 25° followed by 5-minute heatings to 90° are sufficient. If the product does not crystallize when the solution is cooled, seeding can be induced by completely freezing and remelting the solution. On the other hand, several hours of standing at 25° often results in large crystals. The yield is about 85%. The product is identified and its purity relative to U(IV) determined by the absorption spectrum of the solid.[2]

D. HEXABROMOURANATE(1 −) SALTS

$$2[UBr_6]^{2-} + Br_2 \longrightarrow 2[UBr_6]^- + 2Br^-$$

Procedure

1. Tetraethylammonium Hexabromouranate (1 −)

Anhydrous nitromethane (5 ml.) and bromine (3 g.) are added to $[(C_2H_5)_4N]_2[UBr_6]$ (4.5 g.) in a small glass bottle, and the container is quickly closed with a polyethylene-lined cap. The solution is swirled or shaken at room temperature until examination of the very dark solution with a bright light indicates no light-colored uranium(IV) salt remaining. After allowing 5 to 10 minutes for completion of the reaction, the solution is pressure filtered at 25° (lower temperatures result in crystallization of impurities) in the dry-bag as described in Sec. C, subsection 1. The black crystalline compound is washed with cold nitromethane and is dried in a stream of inert gas. The yield is about 65%. *Anal.* Calcd. for $[(C_2H_5)_4N][UBr_6]$: U, 28.1; Br, 56.5. Found: U, 28.2; Br, 55.8. The purity with regard to U(IV) can be determined from the spectrum of the solid.[2] Time required for the preparation is less than one hour.

2. Tetraphenylarsonium Hexabromouranate(1 –)

A 5.6-g. quantity of $[(C_6H_5)_4As]_2[UBr_6]$, 5.6 g. of Br_2, and 5 ml. of nitromethane are used, and the procedure is the same as in the synthesis of $[(C_2H_5)_4N][UBr_6]$. The yield is about 65%. *Anal.* Calcd. for $[(C_6H_5)_4As][UBr_6]$: U, 21.6; Br, 43.6. Found: U, 21.7; Br, 44.3.

E. HEXAFLUOROURANATE(1–) SALTS

Procedure

1. Tetraphenylarsonium Hexafluorouranate(1 –)

$$[UCl_6]^- + 6HF \longrightarrow [UF_6]^- + 6HCl$$

■ **Caution.** *Hydrogen fluoride is toxic and produces painful, slow-healing burns. The reactions should be conducted in an efficient fume hood, and rubber gloves should be worn.*

The procedure for the preparation of $[(C_6H_5)_4As][UCl_6]$ is repeated exactly up through the preparation of $[(C_6H_5)_4As][UCl_6]$ in nitromethane solution. This solution is then added quickly to 15 ml. of cold (0 to – 15°) aqueous 48% HF which is being agitated by a slow sparge of dry inert gas. The product is immediately removed by pressure filtration as described in the preparation of $[(C_2H_5)_4N]$-$[UCl_6]$ using a 15-ml. medium-porosity linear polyethylene fritted funnel. The very pale blue-green product is washed with two small aliquots of 48% HF followed by one very small aliquot of dry acetone (in which the product is rather soluble). The product is dried by passage of the dry inert gas through the filter cake. The entire preparation is performed quickly and with minimum exposure to air in the open fume hood rather than in a gloved bag because it is difficult to observe the operations in polyethylene closely without breathing HF fumes leaking from the bag. The HF-wet compound is not very sensitive to atmospheric moisture, and the filter funnel can be taken into the dry-bag for handling of the dry product. The yield is about 60%. *Anal.* Calcd. for $[(C_6H_5)_4As] [UF_6]$: U, 32.4; F, 15.5. Found: U, 31.9; F, 14.9. Purity of the product with regard to both U(IV) and U(VI) can be determined from the absorption spectrum of the solid. The preparation can be performed in less than an hour.

2. Cesium Hexafluorouranate (1 —)

$$2[UCl_6]^{2-} + Cl_2 \longrightarrow 2[UCl_6]^- + 2Cl^-$$

$$6HF + [UCl_6]^- + Cs^+ \longrightarrow Cs[UF_6] + 6HCl$$

■ **Caution.** *This preparation involves the handling of chlorine solutions under pressure. Because container failure could release the toxic gas with considerable violence, the vessels should be heated in an efficient fume hood with good shielding.*

To 5 g. of $(C_5H_5NH)_2[UCl_6]$ in a 30- to 50-ml. container as described in the preparation of $[(C_2H_5)_4N][UCl_6]$, and precooled to at least 0°, add 5 ml. of nitromethane which has been saturated with chlorine at 0°. The vessel is capped tightly and is heated to 95– 98° for 5 minutes with occasional agitation to redissolve the chlorine gas. After cooling to 0°, the solution is resaturated with chlorine, and the vessel is sealed and again heated to 95–98° for 5 minutes. These treatments are continued until a Cl_2 addition and heating cycle produces no appreciable precipitate on cooling. This usually requires a total of about four Cl_2 additions and heatings. Since it is difficult to saturate the thick slurry that results from the addition of the uranium(IV) salt to 5 ml. of nitromethane, the nitromethane is saturated with Cl_2 at 0° before being added to the $[C_5H_5NH]_2$- $[UCl_6]$. It is important that the first oxidation (heating) be for a long enough time and with sufficient agitation to oxidize and thus dissolve as much of the salt as possible.

Cesium fluoride (3.8 g.) is added to 5 ml. of cold (0°) 48% HF. ■ **Caution.** *HF gas is evolved.* This solution is sparged with dry inert gas at 0°, and the nitromethane solution of U(V) precooled to

$0°$ is quickly added. The precipitate is filtered, washed, and dried as in subsection 1 above. The yield is about 60%. The preparation requires one to two hours. The compound may be identified and its purity determined from its absorption spectrum.[2,8] The pyridinium salt of $[UF_6]^-$ is soluble under these conditions and does not contaminate the product.

Properties

All the salts of the hexahalouranate(1 −) complexes for which preparative procedures are given here, with the exception of $Cs[UF_6]$, are free of U(IV) and U(VI) if the procedures are carried out as indicated. The $Cs[UF_6]$ always appears to contain a small amount of U(IV) as determined by the absorption spectrum of the solid. The amount of U(IV) does not appear to decrease appreciably if the precipitation is carried out at lower temperature or if the precipitation and filtering are done in a dry-bag. Apparently the preparation of $Cs[UF_6]$ by the other published methods also always results in spectrophotometrically detectable amounts of U(IV).[8,9] The salts may contain U(VI) if proper precautions are not taken to avoid moisture in the halogen-oxidation steps because the reaction of the $[UX_6]^-$ with water produces $[UX_6]^{2-}$ and $[UO_2X_4]^{2-}$ and the $[UX_6]^{2-}$ are reoxidized to $[UX_6]^-$ by the halogen. The U(V) hexahalo salts discussed here all appear to be stable if protected from moisture, but the extent to which they must be protected varies considerably. The tetraphenylarsonium salts appear to be much more stable than the cesium and tetraethylammonium salts. Thus, $Cs[UF_6]$ turns to the deep-green color of U(IV) fluoride complexes in a few hours of exposure to laboratory air, whereas $[(C_6H_5)_4As]$-$[UF_6]$ does not change color in the same time. The tetraphenyl-arsonium salts of $[UF_6]^-$, $[UCl_6]^-$, and $[UBr_6]^-$ all appear to be stable indefinitely when simply stored in a desiccator over $Mg(ClO_4)_2$, but the other salts decompose after a few days. All the salts appear to be stable if stored under vacuum.

References

1. J. Selbin and J. D. Ortego, *Chem. Rev.*, **69**, 657 (1969).
2. J. L. Ryan, *J. Inorg. Nucl. Chem.*, **33**, 153 (1971).

3. K. W. Bagnall, D. Brown, and J. G. H. du Preez, *J. Chem. Soc.,* **1964,** 2603.
4. H. J. Sherrill, D. G. Durrett, and J. Selbin, *Inorganic Syntheses,* **15,** 243 (1974).
5. L. G. Morgan, private communication.
6. J. L. Ryan, *Inorganic Syntheses,* **15,** 225 (1974).
7. J. L. Ryan and C. K. Jorgensen, *Mol. Phys.,* **7,** 17 (1963).
8. M. J. Reisfeld and G. A. Crosby, *Inorg. Chem.,* **4,** 65 (1965).
9. R. A. Penneman, G. D. Sturgeon, and L. B. Asprey, *ibid.,* **3,** 126 (1964).

51. PENTACHLORO(TRICHLOROACRYLYL CHLORIDE)-URANIUM(V)

$$2UO_3 + 6Cl_3CCCl{=}CCl_2 \longrightarrow$$

$$2UCl_5 \cdot [OCClCCl{=}CCl_2] + 4C_3Cl_4O + Cl_2$$

Submitted by H. J. SHERRILL,* D. G. DURRET,† and J. SELBIN*
Checked by D. BROWN† and T. L. HALL†

Pentachloro(trichloroacrylyl chloride)uranium(V), $UCl_5(TCAC)$ (in which TCAC is $Cl_2{=}CClCOCl$), is a very useful starting material for the preparation of hexahalouranate(V) salts[1] as well as for other uranium(V) compounds[2] and for compounds of UO^{3+}. We describe here the preparation of this valuable starting compound by a method analogous to the synthesis of uranium(IV) chloride.[3] Elsewhere in this volume, synthetic procedures are given for the preparation of hexahalo complexes of uranium(V), based upon halogen oxidation of the corresponding hexahalouranate(IV) ion.[4]

Procedure

The compounds used in the preparation of $UCl_5 \cdot (TCAC)$ are commercially available and may be used as obtained if they are at least reagent-grade quality.

To 175 ml. (1.24 mole) of hexachloropropene in a 500-ml., three-necked, round-bottomed flask (see Fig. 11 for the complete setup)‡

*Coates Chemical Laboratories, Louisiana State University, Baton Rouge, La. 70803.
†Chemistry Div., Atomic Energy Research Establishment, Harwell, Didcot, Berkshire, England.

‡The checkers obtained satisfactory results by carrying out the reaction with conventional filtration apparatus in a good anaerobic glove box.

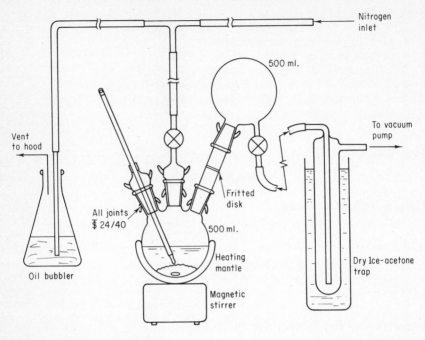

Fig. 11.   Reaction apparatus.

is added 20 g. (0.07 mole) of UO_3 (or U_3O_8). The reaction mixture is heated without stirring by applying approximately 70 volts to the heating mantle. (■ **Caution.** *Various heating mantles respond differently.* The appropriate setting must be determined by trial.) This procedure allows a small amount of the reactants to be heated to a sufficiently high temperature to initiate the reaction without heating the entire mixture to the same temperature. In this way, heating is continued without stirring until the temperature reaches 75°. External heating is stopped when the exothermic reaction starts, as evidenced by the evolution of a white gas and a darkening in the color of the reaction mixture. The temperature continues to rise, and upon reaching 100°, it is maintained between 100 and 120°. Higher temperatures lead to formation of uranium(IV) compounds.

When there remains only a small (*ca.* 0.5 g.) amount of UO_3 in the reaction flask, the solution is filtered by tilting the reaction flask and very gradually applying suction to the receiving flask (see Fig. 11). The gradual application of suction and an increase in the nitrogen flow are necessary precautions to prevent the suck-back of oil and air from the bubbler into the reaction flask. The flask should be tilted carefully so as to disturb the remaining solid UO_3 as little as possible, because the UO_3 is very finely divided and may clog the filter, thereby drastically reducing the rate of filtration. After filtration, the suction is stopped and the pressure is equalized through the sintered-glass filter. If this does not equalize the pressure, the vacuum tubing can be replaced with nitrogen flow to equalize the pressure. Once the filtration is complete, the receiving flask containing the solution is removed from the setup under a nitrogen stream and is stoppered quickly. The filtrate is cooled in an ice bath while a flow of dry nitrogen is bubbled through the solution by using a sintered-glass bubble stick inserted through the appropriately sized rubber stopper in the top of the flask. The gas is allowed to exit through the side arm. A sufficiently high rate of nitrogen flow is necessary to ensure no inlet of moisture or oxygen. The nitrogen stream enhances crystallization by driving off most excess trichloroacryloyl chloride, in which the product is evidently quite soluble.

After approximately 30 minutes, the nitrogen stream is stopped, and the solution is allowed to remain in the ice bath 3 hours. The desired product precipitates as very dark orange-red crystals which vary in color depending upon crystal size. The product flask is transferred into a polyethylene glove bag which has a dry, oxygen-free nitrogen atmosphere. The nitrogen inlet tube is connected to 1 ft. of flexible rubber tubing ending in a one-hole rubber stopper, sized to fit the top of the sintered-glass filter funnel. The product solution is pressure-filtered through the sintered-glass filter (either coarse or medium) with the inlet stream of nitrogen and washed with two 5-ml. portions of dry, deoxygenated CCl_4, in which the product is only slightly soluble. The product is pressure-filtered as dry as possible, and is then vacuum-dried for 3–5 hours. *Anal.* Calcd. for C_3Cl_9OU: U, 39.1; Cl, 29.1 (water-soluble or ionic chloride). Found: U, 40.0; Cl, 29.8.

Properties

The product, $UCl_5 \cdot (TCAC)$, is extremely moisture-sensitive and fumes upon exposure to the atmosphere. However, it appears to be stable indefinitely in a dry, inert atmosphere. It is slightly soluble in

CCl_4 and $CHCl_3$, and is quite soluble in C_6H_6, CS_2, and $SOCl_2$. Disproportionation into uranium(IV) and uranium(VI) occurs rapidly in dimethyl sulfoxide as well as in water, alcohols, and amines. It melts at 147–149° in a sealed tube.

Trichloroacrylyl chloride appears to be such a poor ligand (as would be expected of an acyl halide) that almost any other ligand is able to take its place in the coordination sphere of the uranium(V) entity. Herein seems to lie the utility of the TCAC compound as a starting material. The $UCl_5 \cdot (TCAC)$ complex has been employed as the starting compound for the preparation of a variety of compounds[5] with general formula $UCl_5(L)_x$ and for the preparation of salts of $[UOX_5]^{2-}$ (X = Cl, F).[2] It is also readily converted to $[UCl_6]^-$ salts (*cf.* reference 4) by reaction with alkali metal or tetraalkylammonium chlorides in thionyl chloride solution.[2]

References

1. J. Selbin, J. D. Ortego, and G. Gritzner, *Inorg. Chem.*, **7**, 976 (1968).
2. J. Selbin, C. J. Ballhausen, and D. G. Durrett, *ibid.*, **11**, 510 (1972).
3. J. A. Hermann and J. F. Suttle, *Inorganic Syntheses*, **5**, 143 (1957).
4. J. L. Ryan, *ibid.*, **15**, 235 (1974).
5. J. Selbin, N. Ahmad, and M. J. Pribble, *J. Inorg. Nucl. Chem.*, **32**, 3249 (1970).

INDEX OF CONTRIBUTORS

SUBJECT INDEX

Names used in this cumulative Subject Index for Volumes XI through XV as well as in the text, are based for the most part upon the "Definitive Rules for Nomenclature of Inorganic Chemistry," 1957 Report of the Commission on the Nomenclature of Inorganic Chemistry of the International Union of Pure and Applied Chemistry, Butterworths Scientific Publications, London, 1959; American version, *J. Am. Chem. Soc.*, **82**, 5523–5544 (1960); and the latest revisions [Second Edition (1970) of the Definitive Rules for Nomenclature of Inorganic Chemistry]; also on the Tentative Rules of Organic Chemistry—Section D; and "The Nomenclature of Boron Compounds" [Committee on Inorganic Nomenclature, Division of Inorganic Chemistry, American Chemical Society, published in *Inorganic Chemistry*, **7**, 1945 (1968) as tentative rules following approval by the Council of the ACS]. All of these rules have been approved by the ACS Committee on Nomenclature. Conformity with approved organic usage is also one of the aims of the nomenclature used here.

In line, to some extent, with *Chemical Abstracts* practice, more or less inverted forms are used for many entries, with the substituents or ligands given in alphabetical order (even though they may not be in the text); for example, derivatives of arsine, phosphine, silane, germane, and the like; organic compounds; metal alkyls, aryls, 1,3-diketone and other derivatives and relatively simple specific coordination complexes: *Iron, cyclopentadienyl-* (also as *Ferrocene*); *Cobalt(II), bis(2,4-pentanedionato)-* [instead of *Cobalt(II) acetylacetonate*]. In this way, or by the use of formulas, many entries beginning with numerical prefixes are avoided; thus *Vanadate(III), tetrachloro-*. Numerical and some other prefixes are also avoided by restricting entries to group headings where possible: *Sulfur imides*, with formulas; *Molybdenum carbonyl*, $Mo(CO)_6$; both *Perxenate*, $HXeO_6^{3-}$, and *Xenate(VIII)*, $HXeO_6^{3-}$. In cases where the cation (or anion) is of little or no significance in comparison with the emphasis given to the anion (or cation), one ion has been omitted; e.g., also with less well-known complex anions (or cations): $CsB_{10}H_{12}CH$ is entered only as *Carbaundecaborate(1—), tridecahydro-* (and as $B_{10}CH_{13}^-$ in the Formula Index).

Under general headings such as *Cobalt(III) complexes* and *Ammines*, used for grouping coordination complexes of similar types having names considered unsuitable for individual headings, formulas or names of specific compounds are not usually given. Hence it is imperative to consult the Formula Index for entries for specific complexes.

Boldface type is used to indicate individual preparations described in detail, whether for numbered syntheses or for intermediate products (in the latter case, usually without stating the purpose of the preparation). Group headings, as *Xenon fluorides*, are in lightface type unless all the formulas under them are boldfaced.

251

As in *Chemical Abstracts* indexes, headings that are phrases are alphabetized straight through, letter by letter, not word by word, whereas inverted headings are alphabetized first as far as the comma and then by the inverted part of the name. Stock Roman numerals and Ewens-Bassett Arabic numbers with charges are ignored in alphabetizing unless two or more names are otherwise the same. Footnotes are indicated by *n.* following the page number.

FORMULA INDEX

The Formula Index, as well as the Subject Index, is a cumulative index for Volumes XI through XV. The chief aim of this index, like that of other formula indexes, is to help in locating specific compounds or ions, or even groups of compounds, that might not be easily found in the Subject Index, or in the case of many coordination complexes are to be found only as general entries in the Subject Index. *All* specific compounds, or in some cases ions, with definite formulas (or even a few less definite) are entered in this index or noted under a related compound, whether entered specifically in the Subject Index or not. As in the latter index, **boldface type** is used for formulas of compounds or ions whose preparations are described in detail, in at least one of the references cited for a given formula.

Wherever it seemed best, formulas have been entered in their usual form (i.e., as used in the text) for easy recognition: Si_2H_6, XeO_3, NOBr. However, for the less simple compounds, including coordination complexes, the significant or central atom has been placed first in the formula in order to throw together as many related compounds as possible. This procedure often involves placing the cation last as being of relatively minor interest (e.g., alkali and alkaline earth metals), or dropping it altogether: MnO_4Ba; $Mo(CN)_8K_4 \cdot 2H_2O$; $Co(C_5H_7O_2)_3Na$; $B_{12}H_{12}^{2-}$. Where there may be almost equal interest in two or more parts of a formula, two or more entries have been made: Fe_2O_4Ni and $NiFe_2O_4$; $NH(SO_2F)_2$, $(SO_2F)_2NH$, and $(FSO_2)_2NH$ (halogens other than fluorine are entered only under the other elements or groups in most cases); $(B_{10}CH_{11})_2Ni^{2-}$ and $Ni(B_{10}CH_{11})_2^{2-}$.

Formulas for organic compounds are structural or semistructural so far as feasible: $CH_3COCH(NHCH_3)CH_3$. Consideration has been given to probable interest for inorganic chemists, i.e., any element other than carbon, hydrogen, or oxygen in an organic molecule is given priority in the formula if only one entry is made, or equal rating if more than one entry: only $Co(C_5H_7O_2)_2$, but $AsO(+)\text{-}C_4H_4O_6Na$ and $(+)\text{-}C_4H_4O_6AsONa$. Names are given only where the formula for an organic compound, ligand, or radical may not be self-evident, but not for frequently occurring relatively simple ones like C_5H_5(cyclopentadienyl), $C_5H_7O_2$ (2,4-pentanedionato), C_6H_{11}(cyclohexyl), C_5H_5N(pyridine). A few abbreviations for ligands used in the text are retained here for simplicity and are alphabetized as such, "bipy" for bipyridine, "pn" for 1,2-propanediamine(propylenediamine), "fod" for 1,1,1,2,2,3,3-heptafluoro-7,7-dimethyl-4,6-octanedionato, "thd" for 2,2,6,6-tetramethylheptane-3,5-dionato, "dien" for diethylenetriamine or N-(2-aminoethyl)-1,2-ethanediamine, "chxn" for *trans*-1,2-cyclohexanediamine, "DH" for dimethylglyoximato, and "D" for the dianion, $(CH_3)_2C_2N_2O_2^{2-}$.

Footnotes are indicated by *n*. following the page number.

Ni(C$_8$H$_{12}$)$_2$ Bis(1,5-cyclooctadiene) nickel, **15**:5

NiC$_{12}$H$_{20}$N$_2$O$_2$ Bis(4-methyl-amino-3-penten-2-onato) nickel(II), **11**:74

[Ni(chxn)$_3$]Br$_2$, **14**:61

[Ni(chxn)$_3$]I$_2$, **14**:61

[NiCl$_2$(C$_2$H$_5$OH)$_4$], **13**:158

[NiCl$_2$(C$_4$H$_{10}$O$_2$)] Dichloro(1,2-dimethoxyethane)nickel(II), **13**:160

NiCl$_2$·2H$_2$O, **13**:156

[Ni(dien)$_2$]Br$_2$, **14**:61

[Ni(dien)$_2$]I$_2$, **14**:61

[Ni(en)$_3$]Br$_2$, **14**:61

[Ni(en)$_3$]I$_2$, **14**:61

NiFe$_2$O$_4$, **11**:11

[NiI(NO)], **14**:88

Ni(NO$_2$)$_2$, **13**:203

[Ni{P(C$_6$H$_5$)(OC$_2$H$_5$)$_2$}$_4$], **13**:118

[Ni{P(C$_6$H$_5$)$_3$}$_4$], **13**:124

[Ni{P(OC$_2$H$_5$)$_3$}$_4$], **13**:112

[Ni{P(OC$_6$H$_4$CH$_3$)$_3$}$_3$], **15**:11

[Ni{P(OC$_6$H$_5$)$_3$}$_4$], **13**:108, 116

[Ni(pn)$_3$]Br$_2$, **14**:61

[Ni(pn)$_3$]I$_2$, **14**:61

[NiV$_{13}$O$_{38}$]K$_7$, **15**:108

[Ni$_2$(N$_2$){P(C$_6$H$_{11}$)$_3$}$_4$], **15**:29

O$_2$(AsF$_6$), **14**:39

O$_2$(SbF$_6$), **14**:39

OsCl$_3$(NO)[P(C$_6$H$_5$)$_3$]$_2$, **15**:57

OsHCl(CO)[P(C$_6$H$_5$)$_3$]$_3$, **15**:53

OsH$_2$(CO)[P(C$_6$H$_5$)$_3$]$_3$, **15**:54

OsH$_2$(CO)$_2$[P(C$_6$H$_5$)$_3$]$_2$, **15**:55

OsH$_4$[P(C$_6$H$_5$)$_3$]$_3$, **15**:56

OsO$_2$, **13**:140

[Os$_3$(CO)$_{12}$], **13**:93

PBrF$_2$O, **15**:194

P(CCl$_3$)Cl$_2$, **12**:290

[{P(CH$_2$O)$_3$CCH$_3$}$_2$PdCl$_2$], *cis*-, **11**:109

P(CH$_3$)H$_2$, **11**:124

P(CH$_3$)(C$_6$H$_5$)$_2$, **15**:128

[P(CH$_3$)$_2$C$_3$H$_6$As(CH$_3$)$_2$], **14**:20

['P(CH$_3$)$_2$C$_3$H$_6$N(CH$_3$)$_2$], **14**:21

P(CH$_3$)$_2$(C$_6$H$_5$), **15**:132

P(CH$_3$)$_2$Cl, **15**:191

P(CH$_3$)$_2$H, **11**:126, 157

[P(CH$_3$)$_2$]$^-$, **13**:27; **15**:188

P(CH$_3$)$_2$[Si(CH$_3$)$_3$], **13**:26

[P(CH$_3$)$_2$]$_2$, **13**:30; **14**:14; **15**:187

[P(CH$_3$)$_2$]$_2$C$_3$H$_6$, **14**:17

P(CH$_3$)$_3$, **11**:128; **15**:35

P(CH$_3$)$_3$·BH$_3$, **12**:135

P(CH$_3$)$_3$HI, **11**:128

P(C$_2$H$_5$)$_3$·BH$_3$, **12**:115

P(C$_4$H$_9$)Cl$_2$, **14**:4

P(C$_4$H$_9$)$_2$Cl, **14**:4

P(C$_6$H$_5$)(CH$_3$)$_2$, **15**:132

P(C$_6$H$_5$)(OC$_2$H$_5$)$_2$, **13**:117

P(C$_6$H$_5$)$_2$Na, **13**:28

P(C$_6$H$_5$)$_2$Si(CH$_3$)$_3$, **13**:26

P(C$_6$H$_5$)$_3$·BH$_3$, **12**:113

[P(C$_6$H$_5$)$_4$]Br, **13**:190

P(C$_6$H$_{11}$)$_3$, **15**:39

PCl(CH$_3$)$_2$, **15**:191

PClF$_2$O, **15**:194

PCl$_2$(C$_6$H$_5$)$_3$, **15**:85

PCl$_2$FO, **15**:194

PCl$_3$(C$_6$H$_5$)$_2$, **15**:199

(PCl$_3$)NSO$_2$Cl, **13**:10

PF$_2$H, **12**:281, 283

(PF$_2$)$_2$, **12**:282

(PF$_2$)$_2$O, **12**:281, 285

PF$_4$(CH$_3$), **13**:37

PF$_4$(C$_2$H$_5$), etc., **13**:39

PFe, **14**:176

(PH$_2$)$_2$C$_2$H$_4$, **14**:10

PH$_3$, **11**:124; **14**:1

P(NCO)$_3$, **13**:20

[PN(C$_6$H$_5$)Br]$_3$, **11**:201

POFCl$_2$, **15**:194

POF$_2$Br, **15**:194

POF$_2$Cl, **15**:194

[PO$_2$(OH)(NH$_2$)]K, **13**:25

[PO$_2$(OH)(NH$_2$)](NH$_4$), **13**:23

(PO$_3$NH$_2$)H$_2$, **13**:24

(PO$_4$)$_3$ClSr$_5$, **14**:126

PS(CH$_3$)$_2$Br, **12**:287

P$_2$(CH$_3$)$_4$, **13**:30; **14**:10; **15**:187

P$_2$Si, **14**:173

P$_2$S$_2$(CH$_3$)$_4$, **15**:185

P$_3$N$_3$F$_4$(C$_6$H$_5$)$_2$, **12**:296

P$_3$N$_3$F$_5$(C$_6$H$_5$), **12**:294

PaBr$_4$·4CH$_3$CN, **12**:226

PaBr$_5$·3CH$_3$CN, **12**:227

PaBr$_6$[N(C$_2$H$_5$)$_4$], **12**:230

PaCl$_4$·4CH$_3$CN, **12**:226

PaCl$_6$[N(CH$_3$)$_4$], **12**:230

PbF$_2$, **15**:165

Pb(fod)$_2$, **12**:74

Pb$_4$Br$_6$Se, **14**:171

Pb$_5$I$_6$S$_2$, **14**:171

Pb$_7$Br$_{10}$S$_2$, **14**:171